Memoirs of an Environmental Science Professor

This book shows, through real and current examples from the field of environmental and wetland science, that personal and professional success depends on persistence and a refusal to compromise on "doing the right thing," which for Professor Mitsch, meant saving some of the world's most important ecosystems, as well as educating future researchers and the general public along the way. Case studies described in this book illustrate that persistence pays off, especially when the cause is motivated by something as important as improving our natural environment. They explain clearly that success is not easy, disasters and failures are part of the process, but having goals result in meaningful steps toward it.

Features:

- Emphasizes how it is possible to develop long-term goals and persistence for success both in the academic and environmental world.
- Offers examples set in universities across America and highlights important national wetlands such as the Florida Everglades, the Kankakee River Marshlands in the Great Lakes region, and Ohio's Olentangy River Wetland Park, a Ramsar Wetland of International Importance.
- Speaks to scientists from across the country and the world.
- Discusses chronologically the developments and the achievements of environmental/wetland fields on a global scale.
- Explains how his personal achievements contributed to the growth of wetland and environmental sciences.

Students and professionals in the physical and biological sciences, including chemistry, environmental science, ecological fields, and environmental policy, and especially environmental consultants such as scientists, managers, and engineers, will feel a sense of camaraderie with Professor Mitsch. His long-standing career and devotion to environmental and wetland sciences are an inspiration to all who currently work in the field, aspire to, or simply harbor a sense of appreciation for the natural world and want to learn more about steps that can be taken to manage and protect our planet and the environment.

Applied Ecology and Environmental Management
Series Editor: Steven M. Bartell, University of Tennessee

Global Blue Economy
Analysis, Developments, and Challenges
Md. Nazrul Islam and Steven M. Bartell

Managing Environmental Data
Principles, Techniques, and Best Practices
Gerald A. Burnette

**Environmental Management Handbook, Second Edition –
Six Volume Set**
Edited by Brian D. Fath, Sven Erik Jorgensen

Sustainable Development Indicators
An Exergy-Based Approach
Søren Nors Nielsen

Environmental Management of Marine Ecosystems
Edited by Md. Nazrul Islam, Sven Erik Jorgensen

Surface Modeling
High Accuracy and High Speed Methods
Tian-Xiang Yue

Eco-Cities
A Planning Guide
Edited by Zhifeng Yang

Ecological Processes Handbook
Luca Palmeri, Alberto Barausse, Sven Erik Jorgensen

**Ecotoxicology and Chemistry Applications in Environmental
Management**
Sven Erik Jorgensen

Ecological Forest Management Handbook
Edited by Guy R. Larocque

Handbook of Environmental Engineering
Frank R. Spellman

Memoirs of an Environmental Science Professor
William Mitsch

For more information on this series, please visit: https://www.routledge.com/
Applied-Ecology-and-Environmental-Management/book-series/CRCAPPECOENV

Memoirs of an Environmental Science Professor

William J. Mitsch

CRC Press
Taylor & Francis Group
Boca Raton London New York

CRC Press is an imprint of the
Taylor & Francis Group, an **informa** business

Designed cover image: Cover photo of William Mitsch provided by the Stockholm International Water Institute (SIWI)

First edition published 2024
by CRC Press
6000 Broken Sound Parkway NW, Suite 300, Boca Raton, FL 33487-2742

and by CRC Press
4 Park Square, Milton Park, Abingdon, Oxon, OX14 4RN

CRC Press is an imprint of Taylor & Francis Group, LLC

ISBN: 978-1-032-44929-6 (hbk)
ISBN: 978-1-032-44935-7 (pbk)
ISBN: 978-1-003-37461-9 (ebk)

DOI: 10.1201/9781003374619

Typeset in Garamond
by KnowledgeWorks Global Ltd.

Contents

Foreword I

I have known and admired Professor William J. Mitsch since I first became President of The Ohio State University in 1990. And, therefore, I was pleased when Professor Mitsch asked me to review his memoirs and to write a Foreword. I first met Professor Mitsch shortly after I came to The Ohio State University when he invited me to visit his research efforts and the Olentangy River Wetlands Research Park. I visited and, all of a sudden, he asked me if I would want to go into the marsh. I had on my suit and my bow tie, but Professor Mitsch, ever persuasive, put on big waders and had me do the same and we went into the marshland. That vignette is perhaps the best description I can give you of this remarkable leader in environmental change. Professor Mitsch is not a zealot; rather he is a believer and an evangelist. But, he also recognizes that change is brought about by persuasion, not by anger and frustration. And, that has been his secret sauce: he has been tenacious, but not pugnacious. And, he has been persuasive and persistent, but has always had a sense of humor and even a twinkle in his eye.

This book is about the journey of one person. But, it is also about the realities of the environmental issues with which we are confronted as a nation and a world. It is about the work of one individual, but it is also about the work of many who understand the need to face the reality that we have a responsibility for each other and for the next generation in terms of quality of life and environmental responsibility. It is also about being a clever and thoughtful researcher with immense practical skills that persuade others, not only by the quality of the research, but by touching and seeing. The wetlands project at Ohio State is one example. Until I was in the middle of the wetlands with my waders, I probably would never have given it a second thought as to how all of these intricacies come together in a marshland. The power of persuasion through storytelling and through seeing is the gift that Professor Mitsch has given to all with whom he has worked and to all who have been and will be impacted by his journey.

This book is remarkably readable, it has a joyful tone to it, and, at the same time, it makes a powerful story about the need for advocacy and change in a hyper-changing world.

E. Gordon Gee
President, West Virginia University,
Morgantown, WV
President, The Ohio State University, Ohio,
USA (1990–1997, 2007–2013)

Professor Mitsch pictured with his wife, Ruthmarie, and OSU President E. Gordon Gee, during a trip to Shanghai, China, in June 2010. Professor Mitsch was on his Einstein Fellowship, awarded by the Chinese Academy of Sciences, visiting wetlands and environmental programs in universities all over China. President Gee was there to lead a ribbon cutting ceremony for OSU's new "headquarters" in Shanghai, the goal of which was to recruit undergraduate students throughout all of China.

Foreword II

I began to write this Foreword to the Memoir of Professor William J. Mitsch on World Wetlands Day, February 2, 2023, a day when the wetlands are in the forefront, and thus so is Professor Mitsch, eminent scholar, ecosystem ecologist and pioneer in ecological engineering. World Wetlands Day is set aside to raise awareness and understanding of the critical importance of wetlands to the environment as a whole, and to the current status of the wetlands as one of the world's most threatened ecosystems.

The chapters of Professor Mitsch's book reveal how his work and influence have made a positive difference in the restoration of rivers and wetlands, specific ecological zones, such as the Florida Everglades, environmental science in general, ecological economics, and environmental challenges globally. He shares with us his lifetime of incomparable, innovative, and impactful work, and his tireless advocacy for the appreciation, creation, restoration, and preservation of the wetlands—and some of his challenges in the academic world.

The wetlands are one piece of the immensely complex and integrated puzzle that must be taken very seriously to delay—mitigate—and to stop, today's climate crisis; they have been described as "a linchpin for climate change mitigation." According to Professor Mitsch, the "wetlands are the salvation of the world" and posits that "we need more of these engines of ecological integrity" that purify water, sequester nutrients, preserve fish and wildlife habitats, and serve as a major source for carbon storage.

Wetland degradation and loss have been estimated to range from 21–90% in the United States, but whatever the accurate percentage, it is clear humans are responsible for these losses—by increasing pollutants in agricultural and urban runoff, introducing non-native species, changing water quality, land development, draining for agriculture, damming to form ponds and lakes, diverting flow, and allowing air pollution caused by toxic substances, from marinas, for example. The impact of these assaults on biodiversity, agriculture, fisheries, water and land, and carbon sequestration is immeasurable and here is where we again salute the work of Professor Mitsch.

Professor Mitsch earned a bachelor's degree in mechanical engineering and entered the workforce as a power plant engineer with Ohio and Chicago utility companies. His career goals began to change, however, when he transitioned to the Chicago company's environmental planning staff, spurring him to continue his education at the University of Florida in this direction. It was his PhD work under Dr. H.T. Odum, renowned pioneer of ecosystem ecology, that solidified Professor Mitch's career goals as a systems ecologist, a wetland scientist—a "wetland warrior" as he has been called, and a "wetland hero" in my mind.

Professor Mitsch joined The Ohio State University in 1986 in the School of Natural Resources as Professor and Assistant Director of Research and remained with the University for 27 years before taking on a new challenge in Florida. Ohio State is where I had the privilege of knowing him. It was through his vision for the Olentangy Wetland Research Park in 1991, now the Wilma H. Schiermeier Olentangy River Wetlands Research Park and the 24th Ramsar Wetland of International Importance in the USA, so designated by an intergovernmental treaty that "provides the framework for conversation and wise use of wetlands and their resources."

He created the wetlands design, raised the funds, and managed and directed the academic programs, research and outreach to provide teaching, research and service for wetlands, river science, and ecological engineering. It is a center where one can learn how wetlands function—if the systems can be restored.

The 50-acre research park, with labs, analytical facilities, and spaces for public programming, is a world-renowned contribution to the University and the discipline of ecological engineering. It is a living laboratory, an "ecological treasure." It has been identified as "one of the most productive riverine wetland research laboratories in the world." Principles learned at the Olentangy Park have been applied to river restorations in the U.S. and around the world. Through this work (and beyond), he earned the global respect of environmental scientists and was awarded the 2004 Stockholm Water Prize, a distinction comparable to a Nobel Prize in the field of water.

Professor Mitsch is recognized as a giant among those who have used their careers and persuasive talents to preserve the environment through his firsthand research memorialized in the authorship of more than 600 scientific studies and more than 20 books, including the most highly used textbook, *Wetlands,* now in its fifth edition. Described as the "wetland bible," *Wetlands* has essentially defined the field of wetland science since its first edition in 1986.

Professor Mitsch has trained more than 85 graduate students and 20 post-doctoral fellows for whom he has served as an advisor and mentor; he has collaborated with scientists in many different fields nationally and globally and designed special programs to enlighten and educate the public. Perhaps most important are his students who follow in his footsteps, and take new footsteps beyond what he has accomplished. He has provided opportunities for them to develop their own research agendas that have led to a multitude of varied careers—all capitalizing on their knowledge to make the environment better, and to gain their own notoriety as ecologist engineers, environmentalists, and conservationists. They have had a substantial head start in their careers thanks to their background in the Mitsch lab.

In 2012, Professor Mitsch was recruited to Florida Gulf Coast University, a young university in the Florida State University System, as Professor, Eminent Scholar, and Director of Everglades Wetland Research Park. This new laboratory attracted visiting wetland scientists from around the world under Professor Mitsch's leadership to address large-scale ecosystem restoration. No one could make a bigger difference in leading this park, developing, and preserving its research programs and safeguarding sustained efforts to assure its future advancement.

While Professor Mitsch has worked with local, state, and federal governments and global agencies, companies and communities as an expert advisor and consultant, testifying before the U.S. Congress, his career is that of a consummate academic, an educator and researcher with a passion for his field and a commitment to others who are working to preserve the environment for the future. But, as an academic, he was also confronted by the politics of academe, at both institutions, which he illustrates in several chapters.

There are often obstacles within the institution itself brought about by faculty jealousy, fear, competition, siloed programs, and old thinking, or by external forces that fail to appreciate the extraordinary nature of forward thinking and novel, innovative programs and facilities that reflect unparalleled creativity and, instead, embrace economic opportunities that would replace that which has been created. Many of these obstacles can curtail or thwart ambition and delay progress, but he knew how to prevail to preserve and go forward with what really matters.

It is hard to even begin to comprehend the enormous positive impact that he and his many students, faculty colleagues, and funders have had on the wetlands, a critical ecosystem. One only needs to scan the documentation of his many honors and awards as well as pages of publications and books to understand the importance of his "science-based solutions to critical issues in water resource management and conservation." This delightful and informative book narrated by Professor Mitsch, in the first person, is a very readable story that highlights some of the enormous successes of his career including some of the frustrations that often accompany an ambitious agenda where others may feel competition. I wanted to read more to learn more about this field and his amazing career. I know his memoirs will be a treat to those who have been fortunate to connect with him during his career and celebrate the lifelong, indelible contributions of Professor William J. Mitsch.

Karen A. Holbrook, PhD
Regional Chancellor, University of South Florida
Sarasota-Manatee campus
Former President, The Ohio State University

Professor William J. Mitsch and The Ohio State University President Karen A. Holbrook, honoring the YSI Corporation, an Ohio instrument company, at a social event on April 30, 2007 for a significant donation to the Olentangy River Wetland Research Park, Columbus, Ohio, that included unveiling of the YSI Data Control Center in the new Heffner Wetland Building at the ORWRP.

Preface–Life as a Wetlands Warrior

This book began to formulate in my mind as a nature conservation and restoration book with a working title of "What I Fought For," temporarily borrowing the theme from my undergraduate university—the Fighting Irish of Notre Dame. In the end, I received some of my publishers' advice to tone down the bellicose nature of that title and simply call this my Memoirs of more than 50 years of mixing science and conservation with a strong dose of nature restoration, ecological engineering, and ecological economics—some of the many fields I contributed toward developing. I hope you enjoy the case studies in each of these nine chapters and recognize how they collectively tell an interconnected story of one professor's career.

My working title for this book of "What I Fought For" certainly fit the description that Florida Gulf Coast University (FGCU) gave to me when I moved there in 2012. FGCU sent a reporter up to Columbus, Ohio, to interview me as an incoming eminent scholar and endowed professor for a story that was then published in one of the university's magazines to coincide with my arrival to southwest Florida. I did several hours of interviews with the writer, mostly in Ohio.

I was not aware that the university provided the text and photos to the *Naples Daily News* (NDN) newspaper, too, probably since my lab was based in Naples instead of at the main campus in Fort Myers. But I was fine that they did so.

The *Naples Daily News* published the story on the front page of the Perspective section of its Sunday morning paper of December 30, 2012 (**Figures 0.1 and 0.2**). The NDN is a popular morning read in southwest Florida, especially on weekend mornings by the retirees and wealthy developers who live in the Naples/Fort Myers area. The article was well written and favorable in every way, and I thought it would be well received by the citizens of southwest Florida. I did not realize that the story could be at all provocative.

Figure 0.1 *Photo of Professor William "Bill" Mitsch taken to depict him as a wetland warrior.*

(Photo and permission provided by Brian Tietz. FGCU media and legal affairs also approved use of this photo.)

Figure 0.2 *Wetlands warrior headline in the* Naples Daily News *article, published on Dec. 30, 2012.*

Before the first week of 2013 was over, I was summoned to a meeting by a university administrator, as apparently some developers, university supporters, and others were concerned that this new "wetland warrior" was going to stop all housing and real estate developments in southwest Florida or mess up carefully negotiated mitigation agreements related to business decisions that involved wetland losses.

I protested that a university reporter wrote the story, not me, and I was firmly told: "Well, you didn't have to answer his questions." Suddenly I realized that I was no longer on the *terra firma* of The Ohio State University, a Big Ten University where academic freedom and tenure are givens. I was entering a media twilight zone where little protection or academic freedom seemed to exist—especially since FGCU was one of the first universities in the country where tenure did not exist for even productive senior professors.

As a result of his interviews, the reporter had characterized me as a warrior through my entire academic career. I had not thought about being a wetland warrior before that time, but I warmed up to the idea. Why not? I remain first and foremost an environmental science professor always looking for the truth, no matter how hard it is to find. But I became a little more careful with the policy implications of my science, particularly in the swamps of South Florida. Several of the following nine chapters in this book then reflect how tension between land developers and wetland conservationists and scientists occasionally happened in my 47 years as an environmental science professor. But little did I know how sensitive an issue my proposal of creating a mangrove wetland park in Naples Bay would be among the business community, especially since the accompanying photograph (Fig. 0.1) showed me poised as a threatening guy who looks like he is holding a bazooka (it was really a soil sample corer) over the title of the story—"Wetlands Warrior."

This is a book that describes a professorship where accomplishing something almost always took a major effort and persistence. That is the overall message of the book—to emphasize that nothing worth doing is easy and many times failures accumulated before success happened. That is what happened as we developed the Olentangy River Wetlands on Ohio State University, as we developed the Ohio River Basin Consortium in Louisville, then its feature project the Boatload of Knowledge down the river in the summer of 1987 soon after I arrived at OSU. Defending wetlands from politicians who argued that development was more important in the USA took a major effort in the 1990s, and that priority needs to be protected with every new administration in the federal government.

Protecting the Florida Everglades is a project that will never end. There were significant successes whereby now Indiana and Illinois are both working on restoring wetlands in their dream of someday having an "Everglades of the North" in the former Kankakee Marshlands south of Lake Michigan after they spent a century fighting about whether wetlands should even be allowed to exist.

And yes, it was sometimes a worthy "fight" to develop an interdisciplinary graduate academic program in environmental science or ecological engineering principles in collaboration with Chinese scientists. It was not so much a fight to write a chapter about the "industrial aorta" of the Upper Ohio River Valley where I grew up. It was simply a pleasure for me to tell the story of Appalachia, the Ohio River and how important the region was in the development of our country, especially during the Civil War and World War II industrial productivity. A memoir always gives an opportunity to tell historical notes too.

(Professor Mitsch's Moonlight on the Marsh lecture entitled "What I Fought For," was given on March 24, 2022 at the Everglades Wetland Research Park in Naples, Florida. It can be found on YouTube: https://www.youtube.com/watch?v=CRU7lyft8rE

Acknowledgments

Thank you to West Virginia University President **Dr. Gordon Gee** and University of South Florida Regional Chancellor **Dr. Karen Holbrook** for their kind words in the Forewords they wrote in the front of this book. They remain my two favorite university presidents. I served under both of them at The Ohio State University. Thanks to the many individuals who gave permission for illustrations in this book, including **Brian Tietz** of Fort Myers, Florida, for providing his photo of me as a wetland warrior; **Sean Duffy**, Local History Specialist, Ohio County (WV) Public Library, for providing great historical photos/illustrations of my hometown of Wheeling, Virginia, and West Virginia; **Dr. Lynn Elfner**, editor, *Ohio Journal of Science*, for photos and a journal cover copy from the *Ohio Journal of Science*; and **Judy Kauffeld** for permission to use part of an article she wrote about the Olentangy River Wetland Research Park when it was just an idea.

Thanks to **Dr. Wenshaw He** for a great Chinese wetland photo for the China chapter; **Michelle Drobik**, Reference Archivist at Ohio State University, for tracking down photos featured in Chapter 7. **Dr. Ruthmarie Mitsch** for her strong manuscript editing and assistance with seeking permission for photos; and my brother **Robert L. Mitsch**, for giving permission for an Ohio River photo in Louisville. **Dr. Li Zhang** and **Anne Mischo** had significant roles in these books just as they have had in many previous publications.

Irma Britton and **Chelsea Reeves**, from the publishing houses CRC Press/Taylor & Francis Group, were a pleasure to work with. They clearly know the publishing business the way it should be. Thanks to everyone who helped on this unique publication.

About the Author

Dr. William Mitsch has been an environmental science/engineering professor for 47 years in four U.S. universities (Illinois Institute of Technology, University of Louisville, The Ohio State University, Florida Gulf Coast University). His longest service was 27 years at The Ohio State University, including 20 years as Founding Director of the Olentangy River Wetland Research Park. Dr. Mitsch's awards have included four U.S. Fulbright awards (to Denmark, Botswana, Poland, and Wales), the 2004 Stockholm Water Prize (considered to be the equivalent of a Nobel Prize in water), an Einstein Professorship from the Chinese Academy of Sciences (2010), and a doctorate *honoris causa* from University of Tartu, Estonia (2010).

He was most recently the Director of the Everglades Wetland Research Park in the Water School at Florida Gulf Coast University and Juliet C. Sproul Chair for Southwest Florida Habitat Restoration and Management for 2012–2022.

Significant Honors and Titles:
2004 Stockholm Water Prize Laureate
2010 Einstein Professorship, Chinese Academy of Sciences
2010 Doctorate *honoris causa*, University of Tartu, Estonia
2012 Emeritus Professor, School of Environment and Natural Resources, The Ohio State University
2018 Odum Award for Ecological Engineering Excellence, AEES
2022 Emeritus Professor, The Water School, Florida Gulf Coast University
four U.S. Fulbright awards (to Denmark, Botswana, Poland, and Wales)

Other Significant Awards:
Four Fulbright awards—Denmark, Botswana, Poland, and Wales
U.S. EPA and Environmental Law Institute Wetland Research Award (1996)
Einstein Professorship from the Chinese Academy of Sciences (2010)

Ohio State Professor Emeritus School of the Environment and Natural
    Resources (2012)
Emeritus Professor, The Water School, Florida Gulf Coast University (2022)
Theodore M. Sperry Award for a career in ecological restoration (2005)
Honorary doctorate University of Tartu, Estonia (2010)
Developer of the field of ecological engineering
Advisor to various universities globally

Supporting Interdisciplinary Environmental Science in a Large Midwestern University

Wetland graduate students at the 2005 graduation from The Ohio State University, left to right: Chris Anderson (Ph.D. graduate, School of Natural Resources), Cassandra Tuttle and Amanda Nahlik (M.S. graduates in Environmental Science Graduate Program), and advisor Dr. Bill Mitsch.

(Photo property of and permission from: William J. Mitsch.)

DOI: 10.1201/9781003374619-1

1

1.1 Introduction

In essence, dealing with environmental problems does not fit neatly into the academic structure of universities, especially as we try to solve such issues. A carefully designed "environmental science" or "environmental engineering sciences" program is better. In the late 1980s, American universities were talking the talk but not walking the walk about rigorous but interdisciplinary programs for the benefit of students. Rather, a silo mentality among university departments and schools prevailed, and students, especially graduate students interested in topics such as environmental science and engineering, often had options that narrowed their academic opportunities to the science or profession in the name of the department, which was usually not "environmental science."

When I arrived in the School of Natural Resources at The Ohio State University in January 1986 as a Professor and Assistant Director of Research, I began to notice that our department was not recognized favorably by the "Arts and Sciences" departments that were across the Olentangy River, partially because we did not have a Ph.D. program. I am sure that I had been told in my interviews not to worry about that and that developing such a Ph.D. program would likely be part of my job as Assistant Director of Research.

1.2 Starting an interdisciplinary wetland ecology program at The Ohio State University

An opportunity fell out of the sky a few months after I arrived in Columbus, when the first edition of my textbook *Wetlands*, written by me and by Jim Gosselink, a great co-author from Louisiana State University, was published by Van Nostrand Reinhold in New York City (Mitsch and Gosselink, 1986; **Figure 1.1**).

The book immediately picked up a favorable book review in *Science* (Livingston, 1986), suggesting that wetlands and the field of wetland science were something new and different from the fields of limnology, coastal oceanography, and terrestrial ecology. The *Wetlands* textbook was also received favorably in *American Scientist, Ecology*, and *BioScience*. Wetland science was just emerging as a new interdisciplinary paradigm for universities to consider as well as it was a middle ground between terrestrial ecology and aquatic ecology. For example, many of the ecological theories on aquatic succession needed to be reviewed in the context of whether they were paradigms limited to only aquatic or terrestrial ecosystems but not to the broader concepts of wetlands. Our assessment of how wetlands fit into terrestrial and aquatic successional theories in Mitsch and Gosselink (1986) became one of our most important contributions to this new "wetland science."

Figure 1.1 *Cover of the first edition of* Wetlands, *a textbook by Mitsch and Gosselink (1986) that was published a few months after I arrived at The Ohio State University.*

(Book cover permission from John Wiley & Sons, Hoboken, New Jersey.)

The book and the wetland course that I began to teach annually at Ohio State based on that book must have almost immediately caught the attention of potential grad students interested in wetlands. With that attention, I first began to build a team of talented M.S. grad students. In rapid succession, five graduate students joined my lab and finished their M.S. degrees in Natural Resources, all within my first two years at Ohio State (**Table 1.1**).

The *Wetlands* book, currently in its 5th edition (Mitsch and Gosselink, 2015) and soon to be published in its 6th edition (Mitsch et al., 2024), has sold thousands of copies as the standard textbook for wetland science and ecology courses or as reference books. Of my wetland graduate students listed

Table 1.1 First Five Master's Students Completed by Prof. Bill Mitsch at The Ohio State University, 1986–89

Name	Grad Year	Degree	Thesis Title	Post-Degree
M. Siobhan Fennessy	1988	M.S.	"Reclamation of acid mine drainage using a created wetland: exploring ecological treatment systems"	Continued on for Ph.D. in ESGP at Ohio State
Teresa Cavanaugh	1988	M.S.	"Water quality trends of the upper Ohio River, 1977–1987"	M.S. thesis was paper of the year in 1991 in *Ohio Journal of Science*
Julie Cronk	1989	M.S.	"A model of the Ohio River oil spill of 1988"	Continued on for Ph.D. in ESGP at Ohio State
Kimberly B. Baker	1989	M.S.	"Modelling the fate of iron in a wetland receiving mine drainage"	Became Ohio Department of Natural Resources (DNR) wetland coordinator after completion of M.S. degree
Doreen Robb Vetter	1989	M.S.	"Diked and undiked freshwater coastal marshes of Western Lake Erie"	Employed by USEPA Office of Wetlands, Oceans, and Watersheds, Washington, DC after M.S. degree; currently USEPA Chesapeake Bay Program, Annapolis

in **Appendix 1.A**, many received wetland-related environmental science degrees at The Ohio State University, and are currently teaching wetland ecology courses or even running wetland programs at universities such as Kenyon College (Ohio), Morehead State University (Kentucky), the University of Louisiana, Louisiana State University, Florida Gulf Coast University, Oregon State University, George Mason University (Virginia), Dennison University (Ohio), The Ohio State University, Duke University, Central College (Iowa), and the University of Oklahoma.

1.3 Emergence of a Ph.D. program in Environmental Science at Ohio State

Soon after I arrived at Ohio State, I found an Environmental Biology Ph.D. program "buried" in the Zoology Department's basement across the Olentangy River from my School. The Director of this Ph.D. program allowed professors from other departments to have students in that program, so I jumped on the opportunity and quickly signed up two Ph.D. students and one M.S. student in this Environmental Biology program. One of the doctoral students was Siobhan Fennessy, who had just completed an M.S. under me in 1988. Within four years of my arrival at Ohio State, we were running one of the best research labs in the School of Natural Resources, principally because we had a nice combination of M.S. and Ph.D. students and an occasional post-doc.

Then, the strangest academic shuffle that I have ever seen in my academic life happened—and when I tell this story, nobody can believe it. The folks in the Zoology Department saw that professors without Ph.D. programs were using "their" Environmental Biology program and putting good students in it. The Zoology Department decided then to offer this Ph.D. program to our School of Natural Resources for free as we had no Ph.D. program. I never understood why they were offering this Ph.D. to us. How did the Zoology Department benefit by giving a degree away? Perhaps they thought that the Environmental Biology program was "beneath" the standards of their Zoology program and Zoology Ph.D. Perhaps they felt that we were "using" Zoology Department funds to subsidize the School of Natural Resources, and worse yet, with an interdisciplinary degree program. That would all stop if they got rid of the degree. That offer of a Ph.D. degree by the Zoology Department, which at that moment had two Ph.D. programs, continues to puzzle me.

The Director of The School of Natural Resources decided that we would have a special School faculty meeting to discuss and then vote on whether we should accept this doctoral program into our School that did not have a doctoral degree program. After lively discussion, the School faculty actually voted in favor by a few votes to accept the Ph.D. degree gift; As I recall, the total count was something like 15 to 13 in favor of accepting the doctoral program. The School Director, noting that the word "Biology" was in this Ph.D. program and was therefore a "physical science" degree of little use to the many professors in the School who were in "environmental studies" or "environmental education," announced that the vote was not a strong statement by the faculty in favor of the transfer, so he announced that he would alert the Zoology Department that the School would turn down the offer.

Never in my entire life had I seen an academic program commit academic suicide in such a dramatic way. I had discussions with some of my colleagues who favored our School of Natural Resources taking over the Environmental Biology Ph.D. program, and we discussed the idea of starting a petition for dozens of environmental science professors in many departments at The Ohio State University to sign the petition if they favored the idea of renaming the degree as "Environmental Science" and placing it in some appropriate location in the university (the site we favored was in the Graduate School itself). I recall that we were able to get 50 or more signatures in favor of this idea, none opposing. I was then asked by the Graduate School to organize an ad-hoc committee to discuss the idea of changing this degree to "Environment Science" and perhaps placing it in the Graduate School. We had monthly meetings for the entire academic year and, as I recall, there was one Dean based in the Social Sciences who was able to block this new Environmental Science Ph.D. program until the next academic year. But we persevered and were able to get the transfer approved to convert the "Environmental Biology Ph.D. Program" in Zoology into an "Environmental Science Graduate Program" (ESGP) at The Ohio State University's Graduate School. To this day, the ESGP program remains in place and is still administered through the Graduate School. And the School of Natural Resources would not have another opportunity like this one for having an approved Ph.D. degree for 15 or more years after this unfortunate negative administrative decision by its Director and some of its faculty.

Our Environmental Science Graduate Program at The Ohio State University has supported an interdisciplinary program to this day. I advised 31 ESGP graduate students from 1989–2014 and enjoyed each student's defense working with an interdisciplinary committee (see **Appendix 1.A**). To this day, I remain puzzled by the enormous effort it took to finally get an interdisciplinary Environmental Science degree at Ohio State, and further, by how much energy is even now required every year to protect it from siloed academic departments.

1.4 Principles and current status of OSU's current Environmental Science Graduate Program

The ESGP web page describes a healthy history starting with a few interdisciplinary and bold professors in the late 1980s through today (https://esgp.osu.edu/). As stated on the ESGP web page (as of 2022):

This program is designed around the following principles:

1. Academic compartmentalization is ill-suited to learning about and solving the environmental problems of the 21st century.
2. ESGP is designed to cut across traditional academic disciplines to provide sound and effective graduate education and research on these important environmental issues.
3. Scientists and professionals are needed who, after achieving a solid disciplinary education in science-related fields as undergraduates, have integrated interdisciplinary approaches to solve complex environmental issues.

ESGP's goal is the pursuit and dissemination of knowledge in the interdisciplinary field of environmental science. The program emphasizes basic research on ecological processes and effects, and applied research and teaching that will contribute to solving the world's pressing environmental problems.

Bravo to many environmental scientists and engineers at The Ohio State University for having the guts to support this interdisciplinary environmental science program for now more than 30 years. I was proud to have played a small role in getting it started and having some of my best and most successful graduate students complete their degrees in ESGP.

References

Environmental Science Graduate Program web page at The Ohio State University. 2022. https://esgp.osu.edu/

Livingston, K. 1986. Review of Wetlands, W.J. Mitsch and J.G. Gosselink, eds. *Science* 233 (4769):1209.

Mitsch, W.J. and J.G. Gosselink. 1986. *Wetlands*. Van Nostrand Reinhold, New York, N.Y. 539 pp.

Mitsch, W.J. and J.G. Gosselink. 2015. *Wetlands*, 5th ed. John Wiley & Sons, Inc. Hoboken, NJ., 736 pp.

Mitsch, W.J., J.G. Gosselink, C. Anderson, and S. Fennessy. 2024. *Wetlands*, 6th ed, John Wiley & Sons, Inc., Hoboken, NJ.

Appendix 1.A

Thirty-one graduate students completed by Prof. Bill Mitsch in the interdisciplinary Environmental Science Graduate Program (ESGP) at The Ohio State University over the period 1990 through 2014. Thirteen were M.S. students and 18 were Ph.D. students.

Ph.D. degrees (18)

Jorge Villa Betancur, Environmental Science Graduate Program, The Ohio State University, Spring 2014 "Carbon dynamics of subtropical wetlands communities in South Florida." Current: Assistant Professor, University of Louisiana, Lafayette, LA.

Evan Waletzko, Environmental Science Graduate Program, The Ohio State University, Spring 2014 "Carbon budgets of created riverine wetlands in the Midwestern USA." Current: instructor, College of Arts & Sciences, The Ohio State University.

Kay Stefanik, Environmental Science Graduate Program, The Ohio State University, Spring 2012 "Structure and function of vascular plant communities in created and restored wetlands in Ohio." Current: Iowa Nutrient Research Center, Iowa State University, Ames, IA.

Amanda M. Nahlik, Environmental Science Graduate Program, The Ohio State University, Summer 2009 "Water quality improvement and methane emissions from tropical and temperate wetlands." Current: Research Scientist, Wetland Program, U.S. Environmental Protection Agency, Corvallis OR.

Anne E. Altor, Environmental Science Graduate Program, The Ohio State University, Spring 2007 "Methane and carbon dioxide fluxes in created riparian wetlands in the midwestern USA: Effects of hydrologic pulses, emergent vegetation and hydric soils."

Daniel Fink, Environmental Science Graduate Program, The Ohio State University, Winter 2007 "Effects of a pulsing hydroperiod on a created riparian river diversion wetland." Current: Part-Time Assistant Professor in Zoology, Ohio Wesleyan University, Delaware, OH.

Maria E. Hernandez, Environmental Science Graduate Program, The Ohio State University, Summer 2006 "The effect of hydrologic pulses on nitrogen biogeochemistry in created riparian wetlands in Midwestern USA." Current: Coordinadora de la Red de Manejo Biotecnológico de Recursos, Instituto de Ecologia, Xalapa, Veracruz, Mexico.

Changwoo Ahn, Environmental Science Graduate Program, The Ohio State University, Winter 2001 "Ecological engineering of wetlands with a recycled coal combustion by-product." Current: Professor of Environmental Science, George Mason University, Arlington, VA.

Michael Liptak, Environmental Science Graduate Program, The Ohio State University, Fall 2000 "Water column productivity, calcite precipitation, and phosphorus dynamics in freshwater marshes." Current: Senior Ecologist, EnviroScience, Inc., Stow OH.

Douglas Spieles, Environmental Science Graduate Program, The Ohio State University, Summer 1998 "Nutrient retention and macroinvertebrate community structure in constructed wetlands receiving wastewater and river water." Current: Professor of Environmental Studies, Denison University, OH.

Randall J.F. Bruins, Environmental Science Graduate Program, The Ohio State University, Autumn 1997 "Modeling of flooding response and ecological engineering in an agricultural wetland region of Central China." Current: USEPA, Cincinnati, OH (retired).

Neal Flanagan, Environmental Science Graduate Program, The Ohio State University, Winter 1997 "Comparing ecosystem structure and function of constructed and naturally occurring wetlands: Empirical field indicators and theoretical indices." Current: Visiting Assistant Professor, Environmental Science and Policy Division, Duke University, Durham, NC.

Naiming Wang, Environmental Science Graduate Program, The Ohio State University, Summer 1996 "Modelling phosphorus retention in freshwater wetlands." Current: South Florida Water Management District, West Palm Beach, FL.

Paul Weihe, Environmental Science Graduate Program, The Ohio State University, Summer 1996 "Colonizing and introduced vegetation in created riparian wetlands: establishment during the first two growing seasons." Current: Associate Professor of Biology, Central College, Pella, IA.

Robert Nairn, Environmental Science Graduate Program, The Ohio State University, Summer 1996, "Biogeochemistry of newly created riparian wetlands: evaluation of water quality changes and soil development." Current: David L. Boren Professor and Viersen Presidential Professor, Director, Center for Restoration of Ecosystems and Watersheds, School of Civil Engineering and Environmental Science, University of Oklahoma, Norman, OK.

Julie Kay Cronk, Environmental Science Graduate Program, The Ohio State University, 1992 "Spatial water quality and aquatic metabolism in four newly constructed freshwater riparian wetlands." Current: Professor, Biological and Physical Sciences, Columbus State Community College, Columbus, OH.

M. Siobhan Fennessy, Environmental Biology Program, The Ohio State University, 1991, "Ecosystem development in restored riparian wetlands." Current: Jordan Professor of Biology and Environmental Studies, Kenyon College, Gambier, OH.

Brian C. Reeder, Environmental Biology Program, The Ohio State University, 1990 "Primary productivity, sedimentation, and phosphorus cycling in a Lake Erie coastal wetland." Current: Professor of Biology, Morehead State University, Morehead, KY

M.S. degrees (13)

Yiding Zhang, Environmental Science Graduate Program, The Ohio State University, Spring 2012 "Predicting river aquatic productivity and dissolved oxygen before and after dam removal in Central Ohio, USA." Completed ESGP Ph.D. program, Ohio State University. Current position: Geosyntec, California.

Jackie Batson, Environmental Science Graduate Program, The Ohio State University, Autumn 2010 "Denitrification and a nitrogen budget of created riparian wetlands." First post-degree position: USGS National Research Program for Water Resources, Reston VA.

Kurt S. Keljo, Environmental Science Graduate Program, The Ohio State University, Spring 2009 "Effects of hydrologic pulsing and vegetation on invertebrate communities in wetlands." Current: Franklin Soil and Water District, Columbus, OH.

Cassandra L. Tuttle (Brachfeld), Environmental Science Graduate Program, The Ohio State University, Autumn 2005 "The effects of hydrologic pulsing on aquatic metabolism in created riparian wetlands" Current: Environmental consulting, ARCADIS U.S., Inc., Princeton, NJ.

Amanda M. Nahlik, Environmental Science Graduate Program, The Ohio State University, Autumn 2005 "The effects of river pulsing on sedimentation in two created riparian wetlands." Continued in ESGP Ph.D. program, OSU.

Eric Lohan, Environmental Science Graduate Program, The Ohio State University, Winter 2004 "A methodology to ecologically engineer watersheds for nitrogen nonpoint source pollution control." Current: Consulting in New Mexico

Cheri Higgins, Environmental Science Graduate Program, The Ohio State University, Spring 2002 "Ecosystem engineering by muskrats (*Ondatra zibethicus*) in created freshwater marshes."

Amie Gifford, Environmental Science Graduate Program, The Ohio State University, Winter 2002 "The effect of macrophyte planting on amphibian and fish community use of two created wetland ecosystems in central Ohio."

Dan Fink, Environmental Science Graduate Program, The Ohio State University, Autumn 2001 "Efficacy of a newly created wetland at reducing nutrient loads from agricultural runoff." Continued for ESGP Ph.D., OSU.

Lisa J. Svengsouk, Environmental Science Graduate Program, The Ohio State University, Spring 1998 "First-year response of *Typha latifolia* L. and *Schoenoplectus tabernaemontani* (K.C. Gmel.) Palla to nitrogen and phosphorus additions in experimental mesocosms."

Kathy Metzger, Environmental Science Graduate Program, The Ohio State University, Winter 1997 "Self-design of a fish community in a created riparian freshwater marsh: A simulation model."

Uygar Özesmi, Environmental Science Graduate Program, The Ohio State University, Summer 1996 "A spatial habitat model for the marsh-breeding red-wing blackbird (*Agelaius phoeniceus*) in coastal Lake Erie wetlands." Was a professor and department chair in Turkey.

Courtenay Willis, Environmental Biology Program, The Ohio State University, 1991 "Effects of hydrologic and nutrient variability on emergence and growth of aquatic macrophytes." Went on for Ph.D. at Kent State University; then faculty in Biology Department; then faculty at Youngstown State University; then moved to Holland.

2

Supporting Recognition of Appalachia's Forgotten Superhighway, the Ohio River

A sketch of Upper Ohio River Valley and Wheeling, Virginia, including a newly constructed suspension bridge that carried traffic to Wheeling Island and then westward to Ohio. The sketch is from the top of Wheeling Hill. The date is approximately 1850. The first bridge to span a major river west of the Appalachian Mountains, it linked the eastern and western section of the National Road (currently U.S. Route 40).

(Permission provided by Ohio County Public Library Archives, Wheeling, WV.)

DOI: 10.1201/9781003374619-2

2.1 Introduction

My brothers and sisters and I were born on the banks of the Ohio River in North Wheeling Hospital in Wheeling, West Virginia. I was born on March 29, 1947, and raised in this Ohio River town that started out as Wheeling, Virginia, and became West Virginia during the U.S. Civil War in 1861–65. I am pretty sure the room where I spent my first week on this planet in Wheeling Hospital had a great view of the Ohio River (**Figure 2.1**). Since I was one of seven children in the Woodsdale, Wheeling, West Virginia, Mitsch family (Cathy, Billy, Bobby, Joe, Mary Theresa, Jimmy, and Rita), all of whom were born at the same Wheeling Hospital, the odds are high that our mother, Evelyn Glaser Mitsch, had a splendid view of the Ohio River several times over the period from 1945 to 1958.

My connection to the Ohio River itself was at first indirect. When I was young, we frequently "crossed the Ohio" to visit our grandmother Glaser and many aunts, uncles, and cousins in the Glaser and Robinson families who lived in villages and towns along the Ohio, mostly in Bellaire. But we also had a creek (pronounced "crick") that flowed past our home in the eastern suburb of Woodsdale in Wheeling. It was an indirect tributary of the Ohio. The creek

Figure 2.1 *North Wheeling Hospital, Founded in 1854, on the Ohio River in Wheeling, West Virginia, where Professor Mitsch and his six brothers and sisters were born. The hospital was demolished in 1997 and its operations were moved to a site adjacent to I-70 in the eastern suburbs of Wheeling.*

(Permission provided by Ohio County Public Library Archives, Wheeling, West Virginia.)

was rarely called by its real name of Woods Run that likely gave the suburb its name of Woodsdale. Woods Run was not particularly clean, by public health standards, when we were growing up, but it was the aquatic ecosystem that we played in and explored almost daily. It is probably one of the reasons I became a water resource scientist/ecological engineer.

A ten-minute walk downstream along Woods Run from our home led to Washington's Park, named after Lawrence Augustine Washington, George Washington's full brother. Lawrence lived in Wheeling, Virginia, died there, and is buried at the Washington Family Cemetery in nearby Greggsville.

Before the Civil War, our river town was called Wheeling, Virginia. The secession of West Virginia from Virginia in 1863 was led by the well-developed towns on the eastern or Virginia side of the Ohio River, especially by the citizens of Wheeling, Virginia. This new state would have been a constitutionally illegal state in one sense, as states are not permitted by the U.S. Constitution to split into multiple states, but its legitimacy was proclaimed by President Abraham Lincoln and his cabinet because Virginia had seceded from the Union. When West Virginia broke off from Virginia, it remained a part of the Union that Virginia had rejected. I clearly remember the big celebration we had in Wheeling on the 100th anniversary of West Virginia's statehood in 1963 while I was in high school. West Virginia was truly "the child of the storm" (Wittenberg et al., 2020)—the American Civil War—and the West Virginians of the mid-1800s should be recognized as heroes for their outright rejection of slavery and other state's-rights principles of Virginia and the Confederacy. This is the first of many forgotten contributions of the Ohio River citizens to our nation's well-being.

Woods Run flowed into Wheeling Creek, which was a significant tributary of the Ohio River. We knew our connection with the Ohio whenever the Ohio flooded because there was significant back-flooding of the creek during floods on the Ohio River. When there were floods on the Ohio, we did not have to go to the Ohio shoreline to see them: there was plenty of water backed up on Wheeling Creek and sometimes even on downtown Wheeling streets (**Figure 2.2**).

2.2 Working on the Ohio River after college graduation

My first real job, after receiving my undergraduate degree in mechanical engineering at the University of Notre Dame in 1969, was at a power plant called Cardinal Plant, located 30 miles north of Wheeling and on the Ohio side of the river in a tiny town named Brilliant. I remember I spent a lot of time in the year I worked there helping to repair air pollution control systems at the plant. Few people were aware of environmental issues in 1969, but that all changed in 1970, a red-hot political year that witnessed the first Earth Day as well as the Kent State University massacre. By then, I had moved on to

Figure 2.2 *Ohio River flooding in downtown Wheeling, West Virginia, spring 1964.* (Personal photo permission provided by W.J. Mitsch.)

working for the big electric utility in Chicago, Commonwealth Edison, and out of the Ohio River watershed entirely.

2.3 Getting into an environmental career

There is no question that but for spending a year in the Chicago Loop (the name of Chicago's downtown and business hub) at Commonwealth Edison's corporate headquarters, I may have never become an environmental engineer/scientist. The story is as follows: I was going out for lunch in downtown Chicago on April 22, 1970, when I noted a flurry of activity in Daley Plaza near the Picasso sculpture. I went over to ask some participants what was going on, and I was told they were celebrating the first Earth Day. I asked them what their mission was, or some similar question, and I believe that they said they were trying to stop air and water pollution in Chicago and elsewhere. The discussion led to who they thought were the major polluters, and I think they mentioned my company—Commonwealth Edison—which indeed did have a significant number of fossil-fuel-burning power plants, including one just southwest of the Loop. I kept my cool and my employer's name secret, but I was excited that maybe I could help stop the pollution that Edison was producing from within the company. I went back to my office and started inquiring about what my company was doing programmatically to clean up air and water and if I could be a part of that program. As I recall,

I met a geologist, an aquatic biologist, and an engineer, all of whom were scheduled to be in a new office connected to one of the high-level vice presidents, called an Environmental Affairs Department. I was excited as I could be and asked all three of these individuals if I could join them. They were all friendly and thought it was kind of cool that a young engineer wanted to join them. To my delight, I was invited to be part of the department. They had just hired a new Environmental Affairs department director, Mr. Steve Hastings, who was a professional and a very nice man. All of a sudden, I became aware that I may have propelled my career in the direction of environmental science/engineering. I was excited about that. The only thing that worried me was that I did not know what these environmental fields were since I had been trained as a mechanical/industrial engineer, not an environmental engineer.

There were not many places in the country where one could get environmental training at the time. I applied to graduate schools all over the Eastern USA, as did Ruthmarie (my new wife as of May 28, 1970, five weeks after Earth Day #1). The applications paid off as we both received stipend offers for our respective fields at the University of Florida, which did not sound bad at all during the snowy winter in Chicago when we were applying to grad schools.

We moved to Gainesville, Florida, in August 1971, and became Gators. I was in the Environmental Engineering Sciences Department and Ruthmarie was in the Romance Languages Department. We both eventually received Ph.D's. Mine was under Howard T. Odum, who had just moved to the University of Florida in 1971 on the heels of writing a book that absolutely knocked me off my feet: *Environment, Power, and Society* (Odum, 1971). I not only ended up in a great place to learn about the environment, but I also transformed into less of an engineer and more of a systems ecologist and eventually to a wetland ecologist, since there were few of those anywhere. Being a systems ecologist always allowed me to be unintimidated about venturing into new issues and environmental problem solving. There is more about the Odum connection in future chapters of this book.

2.4 Returning to the Ohio River Valley

I returned to the Ohio River Valley eight years after I left it when, in 1979, I moved to my second faculty position, this time at the University of Louisville in Kentucky; I taught there for six years, from 1979 through 1985.

The Ohio River was referred to as the nation's "industrial aorta" and "the greatest concentration of industry on earth" by *The New York Times* in a feature article on Sunday, November 20, 1955 (Rutter, 1955) because it served as an industrial and commercial bloodline. The river, or its basin, has rarely been recognized for its overwhelming importance in the development of the USA. But environmentally, the Ohio River was a polluted sewer for 100 years of the Industrial Century from the 1870s to the 1970s. Most states that bordered the Ohio River showed little concern for its water quality, because the river was often "owned" by the state across the river. Even the federal government, when the Environmental Protection Agency (EPA) was formed, used the river as a dividing line between regions based in Atlanta (Region 4), Philadelphia (Region 3), and Chicago (Region 5). None of those regional headquarters cities appeared to care much about the Ohio River.

2.5 Starting an Ohio River Basin Consortium

When I joined the University of Louisville in 1979, I noted that the Ohio River was not getting the same attention as other major waterways in the USA in terms of environmental research funding. This was partially because it was not as glamorous as the Chesapeake Bay, the Mississippi River, the Hudson River, and the Great Lakes. But probably most importantly, the Ohio River drained Appalachia. I found out that there was a foundation that was obligated to support some research in the Ohio River so, working with colleagues and grad students, we received funding while at the University of Louisville from the Virginia Environmental Endowment to start a multi-university Ohio River Basin Consortium for Research and Education (ORBCRE), and I continued as its executive director for more than 10 years, even when I moved to The Ohio State University in 1986.

2.6 An international conference in Louisville that changed my career

While at the University of Louisville, I chaired, in 1981, an international ecological modeling conference in Louisville that had been held the previous year in Liège, Belgium. The organizer of the Liège Belgium meeting was Professor Sven-Erik Jørgensen from Copenhagen, Denmark. We titled our Louisville meeting "Energy and Ecological Modelling" and published two books from the conference (Mitsch et al., 1981, 1982). I immediately had the "bug" of book authorship and editorship that has followed me even to this day more than 40 years later.

We featured a riverboat tour of the Ohio River during the conference for the several hundred attendees, many from Europe and Asia. I became fast friends with Sven-Erik Jørgensen during that conference (**Figure 2.3**) and especially on our riverboat cruise. Who would have predicted that 23 years later we would be co-laureates of the Stockholm Water Prize? Meeting and getting to know Sven on the Ohio River begat many visits to Scandinavia and eventually a world-class award from royalty for both of us!

Figure 2.3 *A riverboat cruise on the Ohio River in Louisville, Kentucky, on the* Belle of Louisville *in April 1981 as the setting for the banquet for the "Energy and Ecological Modelling" conference: (a) conference chairs William J. Mitsch, left, and Sven-Erik Jørgensen at the riverboat banquet; (b) Sven's wife Mette Vejlgaard Jørgensen by the stern wheel of the* Belle of Louisville.

(Photo 2.3a credit: Ruthmarie H. Mitsch; Photo 2.3b credit: William J. Mitsch.)

2.7 Our "Boatload of Knowledge" trip down the Ohio River

When I was hired by The Ohio State University a few years later (arriving in Columbus on January 1986), I took the Ohio River Basin Consortium with me since I was its Executive Director. While I continued to champion the Ohio River, I noticed that the state of Ohio and my bosses at Ohio State showed relatively little interest in the consortium or the river itself after which the state is named. Then I had an "aha" moment: Ohio does not care about the Ohio River because Ohio does not own the river. The Ohio River belonged to West Virginia and Kentucky, where they shared their border with Ohio.

Shortly after I arrived at The Ohio State University, I was introduced to Mr. Sherman "Jack" Frost, the most learned water resource historian I have ever met. He was known as the "Water Laureate of Ohio" (Frost and Nichols, 1986). He told me many things about the Ohio River that I had never known, and our collaboration led to co-authoring a paper on the history of the Ohio River (Frost and Mitsch, 1989). He was one of a dozen people who greatly influenced me over the past 50 years, most in universities.

I was especially fascinated about a historic event that Jack Frost told me about that happened 160 years earlier in the winter of 1826–27 on the Ohio River. It was called the "Boatload of Knowledge," a boat trip that was described as being responsible for the transfer of more scientific and ecological knowledge to the "young" American Midwest that any other event during that time (Pitzer, 1989). My mind started spinning with ideas and I decided to use our new linkages with other Ohio River scientists to "redo" in 1987 the 1826–27 Boatload trip from Pittsburgh, Pennsylvania, where the Ohio River begins. We designed our Boatload trip to terminate at my former home-base, Louisville, Kentucky, 600 miles downstream from Pittsburgh. I received a grant to support this "river course" from the Virginia Environmental Endowment and enrolled nine graduate students from several Ohio River Basin universities, three of whom helped with the logistics of our boat classroom, and six of whom performed environmental research projects going down the river. But I also had 18 professors and 33 professionals serving as instructors near their place of work, sometimes getting on and off the boat and lecturing on the boat during its trip downstream. We had a van trailing the boat that would then shuttle these Boatload teachers back upstream to their ground transportation.

It took our 1987 Boatload of Knowledge trip 15 days to navigate the 965 km (600 mile) river trip from Pittsburgh to Louisville (**Figure 2.4**). We received enormous press coverage (11 newspaper articles, dozens of radio and TV interviews) down the river and published a special issue of the *Ohio Journal of Science* (**Figure 2.5**) where we documented the original 1826 Boatload of Knowledge trip as well as our 1987 reenactment trip. We were even featured in a TV spot at halftime during an Ohio State University-West Virginia University football game in Morgantown, West Virginia.

My colleagues and I recalled a riverbank fisherman whom we met on the trip who summarized the uniqueness of our trip by stating that he "had never thought about the river as a teacher" (Mitsch et al., 1989). That idea of rivers as teachers never left me for my entire career.

2.8 A River Keeper Award

I received the 1992 "River Keeper Award" from the Ohio River Basin Consortium for Research and Education that I started for my 10 years as the founder and executive director of the ORBCRE (**Figure 2.6**) and for conceiving and directing the Boatload of Knowledge voyage.

(a)

Figure 2.4 Selected photos from the 600-mile 1987 summer Boatload of Knowledge trip down the Ohio River from Pittsburgh, Pennsylvania, to Louisville, Kentucky, including: (a) students visiting John E. Amos coal-burning electric power plant on the Ohio River in Winfield, West Virginia; (b) boat captain Professor Ralph Taylor from Marshall University lecturing about river mussels in the Ohio River; (c) nine participating graduate students, organizers, boat captain, and others at the end of the journey in Louisville, Kentucky. (Continued)

(Photo 2.4a,b permission provided by Mitsch et al., 1989 and by the *Ohio Journal of Science*; Photo 2.4c provided from personal collection of William J. Mitsch.)

Figure 2.4 *(Continued)*

Figure 2.5 Cover of the special issue of the Ohio Journal of Science *that resulted from our 1987 Boatload of Knowledge journey down the Ohio River (Mitsch and Meserve, 1989).*

(Permission provided by Dr. Lynn Elfner, editor, *Ohio Journal of Science*.)

Figure 2.6 *River Keeper Award presented to Bill Mitsch for a decade of service (1982–92) to the Ohio River as organizer and director of ORBCRE and organizer of the 1987 Boatload of Knowledge voyage down the river.*

(Photo of award plaque provided by W.J. Mitsch.)

2.9 Current status of the Ohio River

The Ohio River is much more appreciated by those who live on the river than before. Buildings now routinely face the river instead of inland. Bridges in cities like Louisville, Cincinnati, and Wheeling are treated with great respect and are often recommissioned as walkways after they have been retired. The suspension bridge in my hometown of Wheeling is a popular tourist visit and the Big Four Bridge in Louisville is popular for both city dwellers and tourists to walk or bike from Louisville, Kentucky, to Indiana and back (**Figure 2.7**).

Water quality and aesthetics are getting much better now in the river, partially because of an inter-state organization named the Ohio River Valley Water Sanitation Commission (ORSANCO) has been around since 1948 and operates monitoring programs throughout the watershed and main stem of the river to check for pollutants and toxins that may threaten fish and wildlife or human uses of the river. ORSANCO also sponsors an annual **Ohio River**

Figure 2.7 *Photo of a modern Ohio River urban shoreline from the Big Four Bridge in Louisville, Kentucky. The Big Four Bridge crosses the river at Louisville, Kentucky, to Jeffersonville, Indiana, and is an enormously popular hiking trail over the river.*

(Photo and permission provided by Robert L. Mitsch, Kettering, Ohio.)

Sweep that involves volunteers from the entire length of the river and from the six states that border the Ohio River—Illinois, Indiana, Ohio, Kentucky, West Virginia, and Pennsylvania—on litter cleanup along the river. The Ohio River Sweep now occurs from March through October to remove the greatest amount of litter possible.

I am certain that efforts like the Boatload of Knowledge that we ran years ago from Pittsburgh to Louisville and the Ohio River Basin Consortium for Research and Education (https://www.ohio.edu/orbcre/) that I helped to develop while I was at Louisville, Kentucky, and Columbus, Ohio, made a difference in the quantity and quality of research in the Ohio River Basin and the education of many students in the region of the importance of our river. That legacy continues through some students who participated in the Boatload of Knowledge adventure and are now professors in the river basin while many new students in the basin have participated in the Ohio River Basin Consortium for Research and Education annual meetings. In the end, these efforts made a measurable impact on researching and university teaching on Appalachia and the Ohio River, my birthplace.

References

Frost, S.L. and W.J. Mitsch. 1989. Resource development and conservation history along the Ohio River. *Ohio Journal of Science* 89: 143–152.

Frost, S.L. and W.S. Nichols. 1986. Ohio Water Firsts. Water Resources Foundation of Ohio, Columbus, OH. 90 pp.

Mitsch, W.J., G.W. Mullins, T.M. Cavanaugh and R. Taylor. 1989. The 1987 "Boatload of Knowledge"—Graduate environmental research and education on the Ohio River. *Ohio Journal of Science* 89: 118–127.

Mitsch, W.J. and L.A. Meserve (eds.). 1989. The Ohio River—Its history and environment. *Special Issue of Ohio Journal of Science* 89: 116–195.

Mitsch, W.J., R.W. Bosserman and J.M. Klopatek (eds.). 1981. Energy and Ecological Modelling. Elsevier Scientific Publishing Co., Amsterdam. 839 pp.

Mitsch, W.J., R.K. Ragade, R.W. Bosserman and J.A. Dillon, Jr. (eds.). 1982. Energetics and Systems. Ann Arbor Press, Ann Arbor, MI. 132 pp.

Odum, H.T. 1971. Environment, Power, and Society. Wiley Interscience, New York, NY.

Pitzer, D. 1989. The original Boatload of Knowledge down the Ohio River: William Maclure's and Robert Owen's transfer of science and education to the Midwest, 1825–1826. *Ohio Journal of Science* 89: 128–142.

Rutter, R. 1955. The Ohio becomes industrial aorta. *The New York Times*, Section 3, November 20, 1955. pages 1, 8.

Wittenberg, E.J., E.A. Sargus, Jr. and P.L. Barrick. 2020. Seceding from Secession: The Civil War, Politics, and the Creation of West Virginia. Savas Beatie, El Dorado Hills, CA. 268 pp.

3

Using Ecological Economics to Settle a Civil War between Indiana and Illinois on the Kankakee River Marshlands

Kankakee River marshland and floodplain near Momence, Illinois, immediately downstream from the Indiana State line. Boat ride on the Kankakee River near Momence, Illinois.

DOI: 10.1201/9781003374619-3

3.1 The Great Kankakee Marshlands

In the 19th century, the Great Kankakee Marshlands was one of the largest marsh-swamp basins in the interior United States. Located primarily in northwestern Indiana and northeastern Illinois, the Kankakee River basin was 13,700 km² in size, with 8,100 km² located in Indiana, the site of the majority of the original Kankakee Marsh (**Figure 3.1**). From the river's source near South Bend, Indiana, to the Illinois state line, a direct distance of only 120 km, the river originally meandered through 2,000 bends along 390 km, with a nearly level fall of only 8 cm per km. Numerous wetlands, primarily wet prairies and marshes, were virtually undisturbed until the 1830s, when settlers began to enter the region. The naturalist Charles Bartlett (1904) described the wetland as follows:

> *More than a million acres of swaying reeds, fluttering flags, clumps of wild rice, thick-crowding lily pads, soft beds of cool green mosses, shimmering ponds and black mire and trembling bogs—such is Kankakee Land.*

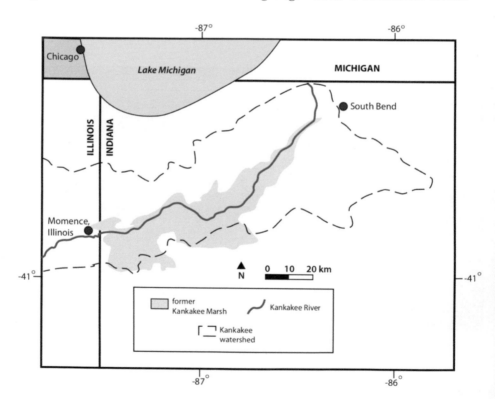

Figure 3.1 *Original area of the Great Kankakee Marsh and its watershed in north-western Indiana and northeastern Illinois. Location of Momence, Illinois, is also shown. The current Momence wetland that is described below is between the town of Momence and the Illinois-Indiana state line.*

(Original sketch, property of W. J. Mitsch; permission provided by W. J. Mitsch.)

These wonderful fens, or marshes, together with their wide-reaching lateral extensions, spread themselves over an area far greater than that of the Dismal Swamp of Virginia and North Carolina.

Bartlett, 1904

The Kankakee region was, for thousands of years, a prime hunting ground for Native Americans. An estimated 400 generations of Native Americans lived off the marsh.

When the early French explorers, including LaSalle, first visited the Grand Kankakee Marsh in 1679, it was a million-acre wetland that "was as wide as it was long" (Indiana Everglades, 2009). It remained pristine for more than 150 years until the drainage began.

After the Civil War, the wetland became a sportsman's paradise where "water-fowl blackened the skies." For a short time, it was considered a prime hunting area for European settlers, until wholesale draining of the land for crops and pasture began in the 1850s and continued especially after the Civil War. Singleton Ditch was dug 3 m (10 feet) deep and 150 km (90 miles) long for the purpose of draining these wetlands. The impact was enormous on one of the largest inland marshes in North America: It was estimated that this wetland drainage alone decreased by 25% of the number of birds in the Mississippi flyway.

The Kankakee River and most of its tributaries in Indiana were channelized into straight ditches in the late 19th century and early 20th century. In 1938, the Kankakee River in Indiana was reported to be one of the largest drainage ditches in the United States; the Great Kankakee Marsh was essentially gone by then. The Great Kankakee Marsh, which was located less than 50 miles south of the current Chicago urban sprawl, was a result of the Wisconsin glacier that began 70,000 years ago and existed until about 3,000 years ago. Over many thousands of years, the glacier moved soil and rock to the south, creating the U-shaped Valparaiso moraine that encircles the southern shoreline of the current Lake Michigan. Outwash along the Kankakee River, which extends from South Bend to the Indiana-Illinois border, is 24 to 32 km (15 to 20 miles) wide, including much of the original Kankakee Marsh. A ridge of limestone at Momence, Illinois, formed a natural dam on the Kankakee River. Before its draining, the Kankakee Marsh was one of the largest freshwater marshes in the United States (Wikipedia, 2021) and still today is referred to by many conservationists as the "Everglades of the North" (see Everglades of the North, 2012).

3.2 Our study estimating the value of Kankakee River Wetlands

I was contacted by the state of Illinois to provide estimates of the values, in dollars, if possible, of a small section of the former Kankakee Marshlands called the Momence Wetlands (**Figures 3.2** and **3.3**), which remained in Illinois.

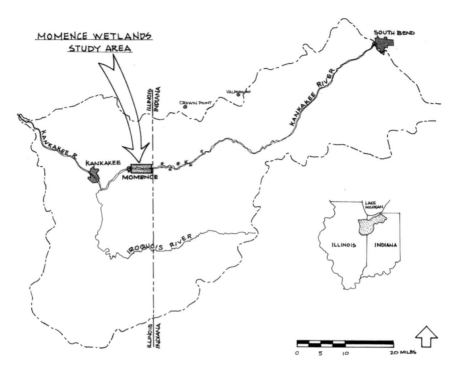

Figure 3.2 Kankakee River watershed showing location of the Momence wetlands study site.

(Unpublished sketch property of William J. Mitsch.)

The state of Indiana was applying pressure on Illinois to consider draining the Momence Wetland by removing the limestone outcrop at Momence that would significantly lower the water level both in the Momence Wetland but also in the drainage ditches in Indiana. The entire site shown in **Figure 3.3** is 1,896 hectares (4,683 acres), with 769 hectares (1,899 acres) classified as wetlands in three categories (**Table 3.1**).

More than 40 years ago, in 1978–79, Mr. George Benda was the Project Officer for the grant I received from the Illinois Institute of Natural Resources (IINR) office in Chicago. I had been invited by Benda and the IINR to provide estimates of the value of riverine wetlands along the Kankakee River in northeastern Illinois to support their legal battle with the State of Indiana on the management of a remnant of the Great Kankakee Marshlands and River. I was in my fourth year at Illinois Institute of Technology in Chicago when this project started. And, for all practical purposes, the Great Kankakee Marshlands had already been mostly drained more than a century before. The question we tackled in this project: Can we calculate the economic and energy "values" of these marshlands in Illinois' portion of the Great Kankakee Marshlands to counter the demands from the state of Indiana that the Illinois wetlands be drained to enhance agricultural productivity in downstream Indiana?

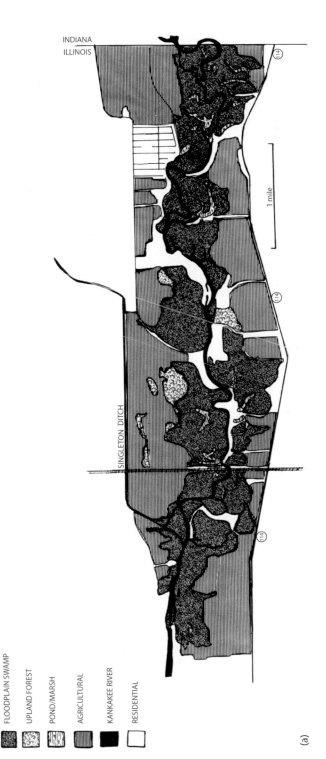

Figure 3.3 Momence Wetlands study area showing. (*a*) present (in 1970s) ecosystems and land use (Continued) (Unpublished sketch property of William J. Mitsch.)

FLOODPLAIN SWAMP

UPLAND FOREST

POND/MARSH

AGRICULTURAL

KANKAKEE RIVER

RESIDENTIAL

(a)

INDIANA
ILLINOIS

SINGLETON DITCH

1 mile

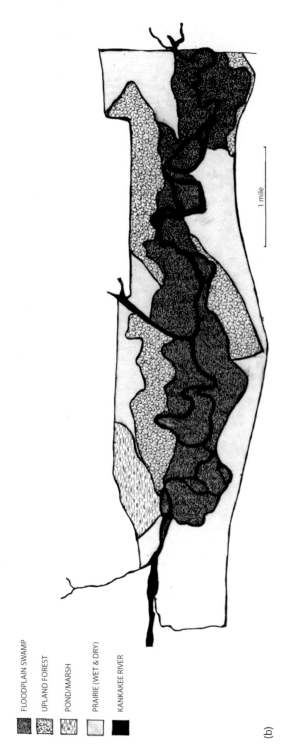

FLOODPLAIN SWAMP

UPLAND FOREST

POND/MARSH

PRAIRIE (WET & DRY)

KANKAKEE RIVER

1 mile

(b)

Figure 3.3 (*Continued*) (*b*) *Original ecosystems in pre-settlement times.*

Table 3.1 Area of Wetlands and Other Land Uses in the Momence Wetlands Study Area Shown in **Figure 3.3**

Ecosystem or Land Use	Acres	Hectares
Floodplain forest*	1,649	668
Upland forest*	128	52
Shrub swamp/Pond*	122	49
Agriculture	2,048	829
Residential	736	298

*wetland and wetland buffer = 769 hectares (1,899 acres).

It is interesting that the same Mr. George Benda who was our project officer in 1979, wrote a fictional book featuring this civil war 38 years later (Benda, 2017). The book was called simply *The River* with renamed characters representing some of us from the real-life 1979 study. I was named "Mitsch Hutchinson" in Benda's book and I was described as a young hot-shot ecologist from University of Chicago with no mention of my real university at the time of Illinois Institute of Technology which is located 20 blocks north of the University of Chicago. It was hard for me to split fact from fiction the first time I read *The River*. Benda's book also stimulated me to write this chapter to give the real story as best I could remember it, see **Figures 3.3 (a) and (b)** above.

3.2.1 Momence wetland functions and values

The 1,899 acres of wetlands at Momence offer a wide variety of fish and game. There are largemouth bass, smallmouth bass, walleye, and carp in abundance, and, at the time of our study, the state record walleye, carp, and smallmouth bass were caught in the Kankakee River. The seasonally flooded floodplain forested wetlands in the Momence region provide vital spawning and feeding grounds for fish during the flooding season (**Figures 3.4 and 3.5**).

The seasonally flooded floodplain forests also enhance water quality by providing sites for sedimentation and phosphorus retention as well as seasonally flooded soils that contribute to nitrate-nitrogen retention. The Illinois State Water Survey estimated that there is a net sedimentation of 104,000 tons of sediment along a 24-km (15-mile) stretch of the river in the Momence area over 10 years. A detailed study by my IIT lab showed that 3% of phosphorus-rich sediments that passed over a riparian wetland in southern Illinois were retained by sedimentation into the wetland (Mitsch et al., 1979c). The trees provide shading of the river and thus decrease average water temperature in the summer months. The floodplain forests provide excellent feeding and

Figure 3.4 *Photographs of riverine floodplain forest in Momence Wetlands* ***(a)*** *in dry season and* ***(b)*** *during flooding season.*

(Photos taken by and property of William J. Mitsch.)

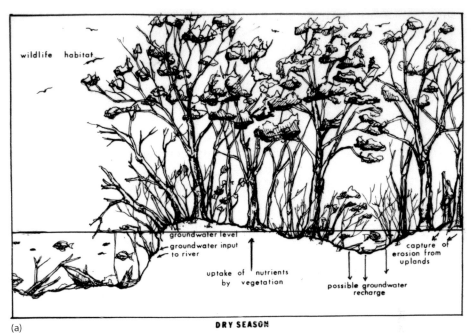

wildlife habitat

groundwater level
groundwater input
to river

capture of
erosion from
uplands

uptake of nutrients
by vegetation

possible groundwater
recharge

(a) **DRY SEASON**

riverbank
stabilized by
vegetation

contribution of
organic food to stream

sediment deposited
from river

storage of floodwater

fish spawning & feeding

nitrate nitrogen
removal from
water

(b) **FLOOD SEASON**

Figure 3.5 Sketch of ecosystem services of Kankakee riverine wetlands during *(a)* flood season and *(b)* dry season.

(From John Wiley & Sons, Inc. *Wetlands*, 2nd ed. 1993 Fig. 15.6, page 520. With permission for both illustrations.)

nesting habitat for avian fauna. Even dead and dying trees provide good habitat for many birds, including woodpeckers. Additionally, the floodplain wetlands provide valuable flood control for human communities along or near the Kankakee River, see **Figures 3.5 (a) and (b)** above.

By the middle of the 1970s, estimating the economic values of ecosystems was a common activity. Ecosystem services were understood, though they were generally called ecosystem values. But the journal *Ecological Economics* was still a few years in the future and conferences on the subject were rare. Metrics other than money such as energy flow were emerging from H.T. Odum's lab at the University of Florida (Odum, 1971; Gilliland, 1975; Odum and Odum, 1976). Westman (1977) posed the question in *Science*: "How much are nature's services worth?"

To answer that question, we used two distinctly different approaches for estimating the value of the Momence wetlands on the Kankakee River: (a) Replacement valuation (b) Energy analysis.

3.2.2 Replacement valuation

Replacement valuation was just becoming an acceptable economic approach in the 1970s for estimating the values of ecosystem services, but it was not more globally accepted until the 1990s. If one could calculate the cheapest way of replacing various services performed by a wetland and could make the case that those services would have to be replaced in the event that the wetland was destroyed, that figure would be the "replacement value."

Some of the replacement technologies that might be necessary to replace services provided by the Kankakee wetland processes are listed in **Table 3.2**. Calculation of the replacement cost method is illustrated in Mitsch et al., 1979a,b. In this example, a fish hatchery is used to calculate fishery production, a flood reservoir to calculate flood and drought control, sediment dredging to estimate sediment retention, and wastewater treatment to estimate water quality enhancement. This approach has the merit of being accepted by some conventional economists. For certain functions, it gives very high values compared with those of other valuation approaches discussed in this section. For example, the tertiary treatment of wastewater is extremely expensive, as is the cost of replacing the nursery function of marshes for juvenile fish and shellfish. Serious questions have been raised about whether these functions would be replaced by treatment plants and fish nurseries if the wetlands were destroyed. Some ecologists and economists argue that in the long run, either the services of wetlands would have to be replaced or the quality of human life would deteriorate.

Table 3.2 Estimated Value of 770-ha (1,900-acre) Riparian Wetlands along the Kankakee River, Northeastern Illinois, Estimated by Replacement Value Approach and by Energy Analysis

Replacement Cost Approach

Ecosystem Function (Replacement Technology)	$/Year	Total Value
Fish productivity (Fish hatchery)	$91,000	
Flood control/Drought prevention (Flood control reservoir)	$691,000	
Sediment control (Sediment dredging)	$100,000	
Water quality enhancement (Wastewater treatment)	$57,000	
Total replacement cost	$939,000	
Value/Area $939,000 yr^{-1}/770 ha =		$1,219 hectare^{-1} yr^{-1} ($ $487 acre^{-1} yr^{-1})

Energy Flow Approach

Energy Flow Parameter	Number	Total Value
Ecosystem gross primary productivity (kcal m^{-2} yr^{-1})	20,000	
Energy quality conversion (kcal GPP/kcal fossil fuel)	20	
Energy conversion in U.S. economy (kcal fossil fuel /U.S.$)	14,000	
Value/Area =		$714 hectare^{-1} yr^{-1} ($285 acre^{-1} yr^{-1})

(From Mitsch et al., 1979a; Mitsch and Gosselink, 2015)

3.2.3 Energy analysis

A simple calculation using the annual energy flow of a bottomland forested wetland in Illinois is illustrated in the bottom half of **Table 3.2**. Here, an estimated ecosystem energy flow (gross primary productivity = GPP) of 20,000 kcal m^{-2}yr^{-1} yielded an estimated value of $714 ha^{-1}yr^{-1} ($285 acre^{-1} yr^{-1}). The energy analysis method gave a number about 60 percent of the replacement value. The concept of energy "quality" was used in this calculation to differentiate between energy flow in the ecosystem (based on gross primary productivity) and energy flow in the human-based fossil fuel economy. This method is a precursor to a more recent approach of emergy developed by H.T. Odum.

Regardless of which kind of ecosystem evaluation is used, several generic problems and paradoxes in quantifying wetland values were pointed out by us (Mitsch and Gosselink, 2015). If one ignores the technical problems of functional ecosystem substitution, the idea attracts many people because of the common perception among economists that any commodity can be replaced. They believe that as scarcity of one product drives the price up, the creativity of the free market will surely result in the development of a cheaper substitute. Yet, this is not true of ecosystems. Much of the value of an ecosystem, especially an open system such as a wetland, depends on its landscape context and the strong interactions among the parts of the landscape. Thus, the value of a riparian forest depends on its ecological links to the adjacent stream on one side and the upland fields or forest on the other.

Mitsch and Gosselink (2015) were influenced by this economic evaluation of wetlands to publish eight "concerns" about taking economic valuation of ecosystems too seriously:

1. The terms "value" and "service" are anthropocentric; hence, assigning values to different natural processes usually reflects human perceptions and needs rather than intrinsic ecological processes.
2. The most valuable products of wetlands are public amenities that have no commercial value for the private wetland owner.
3. The ecological value, but not necessarily the economic value, of a wetland depends on its context in the landscape.
4. The relationships among wetland area, surrounding human population, and marginal value are complex.
5. Commercial values are finite, whereas wetlands provide values in perpetuity.
6. A comparison of economic short-term gains with wetland value in the long term is often not appropriate.
7. Estimates of values and services, by their nature, are colored by the biases of individuals and society and by the economic system:
8. A landscape view of wetlands is required to make intelligent decisions about the values of created and managed wetlands.

3.3 Continuing efforts to restore the Kankakee Wetlands

More recently, there has been some effort to restore parts of the Great Kankakee Marsh in northwestern Indiana and even to restore it to the point where it could be a National Park. The idea is attractive, as approximately 10 million Americans are within a two-hour drive of the Kankakee and because there is not a river national park, despite many candidates. A restored Grand Kankakee River could be a one of the greatest national parks. John Hodson of the Kankakee Valley Historical Society (Indiana Everglades, 2009) described the history of the Kankakee River drainage and opportunities for its restoration:

> It's so much bigger than me. And it's such a great story.... And what it does is it teaches us about ourselves. It teaches us about the awe and wonder of the wildlife area and it also teaches us about greed. When the

*land speculators came in [they]realized that if they drained this river ...
they could sell it and make a fortune. And then the realization that we
destroyed something that was so beautiful and it gave the opportunity to
other men to come together and [figure out] how can we fix this machine
that we broke? Because a marsh (and an ecosystem) is a machine. And
we broke it.*

Indiana Everglades, 2009

3.4 How our ecological economic research 40 years ago stopped a civil war

It is not an exaggeration that Indiana and Illinois were in a civil war over the management of the Kankakee Marshlands for almost a century. In Indiana, the wetland was viewed as an impediment to agricultural productivity in upstream Indiana. According to the state of Indiana, the limestone outcrop at Momence needed to go. But attitudes were starting to change in Illinois whose mostly urban populations were becoming more appreciative of the importance of wetlands. When it became known that a professor from Illinois had calculated that the natural wetlands had economic value for both Hoosiers and Illini, and especially when our report (Mitsch et al., 1979a) was published, a recognition of wetland values began to develop in urban Illinois and maybe a little slower in rural Indiana. I was quite pleased even 10 to 20 years ago to see that both states had programs for restoring what they proudly call their "Indiana Everglades."

The only significant obstacle to restoring the marshlands really was the agricultural interests in Indiana. After our report was published, I think appreciation of the importance of wetlands reduced that obstacle where, in less than a generation, local communities in both states came to appreciate wetlands. When George Benda contacted me a few years ago and told me that his fictional book *The River* was published in 2017, it also made some people proud of "their "Indiana Everglades." Someday there will be a National River Park here, perhaps similar to the Florida Everglades, especially since the 10 million potential tourists live within 2 hours of this site. Chicagolanders will love their restored *Kankakee River and Marshland Park.*

References

Bartlett, C. H. 1904. Tale of Kankakee Land. Charles Scribner and Sons, New York, New York, 232 pp.
Benda, George. 2017. *The River.* A Jack Slack Shoebox Dialogue. ISBN: 978-1-54391-275-32012.
Everglades of the North. 2012. IMDb, documentary, https://www.imdb.com/title/tt2522274/, 57 min.
Gilliland, M.W. 1975. Energy analysis and public policy. *Science* 189: 1051–1056.

Indiana Everglades (mp4). 2009. https://www.youtube.com/watch?v=kDnsy5wrGdA. 13 minutes. For Goodness Sakes Productions, LLC.

Mitsch, W.J., M.D. Hutchison and G.A. Paulson. 1979a. Momence wetlands of the Kankakee River in Illinois—an assessment of their value. Illinois Institute of Natural Resources Doc. 79/17, Chicago, IL. 55 pp.

Mitsch, W.J., W. Rust, A. Behnke, and L. Lai. 1979b. Environmental observations of a riparian ecosystem during flood season. Illinois Water Resources Center Research Report No. 142, Urbana, IL. 64 pp.

Mitsch, W.J., C.L. Dorge and J.W. Wiemhoff. 1979c. Ecosystem dynamics and a phosphorus budget of an alluvial cypress swamp in southern Illinois. *Ecology* 60:1116–1124.

Odum, H.T., 1971. Environment Power and Society. John Wiley & Sons, Inc. New York, NY.

Odum, H.T. and E. C. Odum. 1976. Energy Basis of Man and Nature. McGraw Hill, New York, NY.

Westman, W.E. 1977. How much are nature's services worth? *Science* 197: 960–964.

Wikipedia. 2021. Kankakee Outwash Plain. https://en.wikipedia.org/wiki/Kankakee_Outwash_Plain

4

Developing the Principles of Ecological Engineering and Water Management with China

XiXi National Wetland Park, Hangzhou, China.

(Permission provided by W. He.)

DOI: 10.1201/9781003374619-4

4.1 Introduction

Some of my activities involving Chinese researchers over a 30-year period are listed in **Table 4.1** and locations are shown in **Figure 4.1**. Early in our exchanges, we collaborated with several labs associated with the Chinese Academy of Sciences (referred to in early publications as Academia Sinica).

In the 1970s, there was a swing going on in the field of ecology, and also in engineering, that led us at The Ohio State University to secure a collaborative research project, funded on our side by the National Science Foundation (NSF), on a new idea of ecological engineering that was starting to bloom in both the East and the West. Ecological engineering was co-evolving at that time in the USA, Europe, and China. The focus of our NSF grant in the USA was to compare these separately evolving fields with the same name and look for common ground. In fact, I gave the keynote address at the first conference on ecological engineering ever held, on

Table 4.1 Some Collaborations of William Mitsch with Chinese Scientists, Academies, and Universities, 1987–2018

August 2018—Plenary Lecturer, Society of Wetland Scientist Asian Chapter meeting, Northeast Institute of Geography and Agroecology, Chinese Academy of Sciences, Changchun, China

September 2016—Plenary Lecturer, 10th INTECOL Wetlands Conference "Healthy Wetlands, Healthy Earth," Changshu, China

March 2016—Visiting Lecturer, Tsinghua University, Beijing, China with other lectures in Beijing

July 2014—Lecture/wetland tour: Beijing Normal University, Beihang University (Beijing), China Institute of Water Resources and Hydropower Research (Beijing), Northeast Normal University (Changchun), Qufu Normal University (Qufu), First Institute of Oceanography, National Oceanic Administration (Qingdao)

July 2013—Lecture/wetland tour: Chinese Academy of Sciences (Beijing), Northeast Normal University (Changchun), Tsinghua University (Beijing), Sichuan University (Chengdu), Qinghai Normal University (Xining) and wetland tours of Jiling and Qinhai provinces

2011 and 2012—Lectures in several conferences in Yangtze River Basin: Jingzhou, Hubei Province, China (2012) and Chongqing Normal University, Chongqing, China (2011), Wuhan University, Wuhan, China (2012)

2010—Einstein Professorship lecture tour: Shanghai, Haikou, Beijing, Zhengzhou, Chongqing, Wuhan, Guilin, Hong Kong

Short courses, workshops, research collaboration, and conferences at Chinese Academy of Sciences (CAS) and several universities: Suzhou, Beijing, Nanjing, Wuhan, Chongqing, Huaibei, Hefei, and Shanghai with significant extensive visits to China in 1987, 1989, 1996, 2002, 2004, 2005, 2006, 2007, 2008, 2009, 2010, 2011, 2012, 2013, 2014, 2016, and 2018

1987–1996—Worked with CAS scientists in development of the field of ecological engineering

1987–2010—Collaborated with Chinese scientists on Yangtze River/Three Gorges Dam, ecological engineering, carbon sequestration, and wetland ecology

1989–2023—Frequent host to visiting Chinese professors and graduate students at Olentangy River Wetland Research Park, The Ohio State University and Everglades Wetland Research Park, FGCU, Naples, Florida

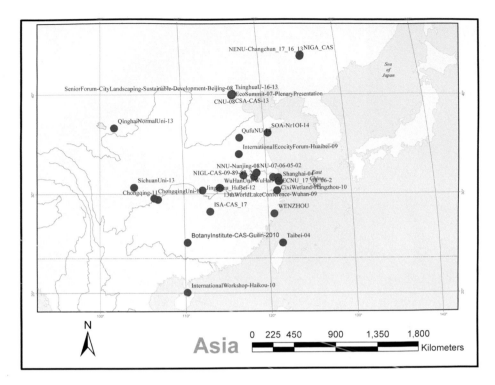

Figure 4.1 *Map of locations in China where Bill Mitsch was invited for presentations and collaborations from 1987 through 2018.*

(Original map prepared for this publication.)

March 24, 1991, entitled "Ecological engineering—roots and rationale for a new ecological paradigm" at an International Conference on Ecological Engineering of Wastewater, in Trosa, Sweden (the paper was finally published as Mitsch, 1987), two months before our team traveled to China to initiate our grant. After my first visit to China in 1987 with Danish Professor Sven-Erik Jørgensen, the two of us collaborated on the first textbook on ecological engineering (Mitsch and Jørgensen, 1989). In that book and in a parallel chapter in a new book on ecological economics by Robert Costanza (Costanza, 1991), we compared approaches to ecological engineering in the West and in China (Mitsch, 1991).

Also, a group of Chinese scientists, headed up by Professor Ma Shijun, visited us in Ohio in early 1989 (**Figure 4.2**), and we reciprocated with a visit of our USA team of researchers in May 1989 (**Figure 4.3**). The first visit by Chinese scientists to The Ohio State University was quite productive, and our visitors were able to travel to sites on Lake Erie and the Ohio River despite the fact that both tours were in winter.

Figure 4.2 *Three visiting scientists from China were treated to a "Chinese banquet" at Bill Mitsch's home in Columbus, Ohio, February 1989. Seated, left to right, Bill Mitsch; Ma Shijun, Center for Eco-environmental Studies, Academia Sinica, Beijing, China; and Li Dianmo, Academia Sinica, Beijing, China. Standing: Bill's daughters, Rebecca Mitsch and Jane Mitsch, and Yan Jingsong, Nanjing Institute of Geography and Limnology, Academia Sinica, Nanjing, China.*

(Permission provided by Ruthmarie H. Mitsch.)

4.2 Ecological engineering for treatment of riverine wastewater with water hyacinths

Our early interactions with Chinese scientists in the late 1990s involved comparison of ecological engineering case studies in the USA and the West and ecological engineering as described by Professor Ma Shijun from the Center for Eco-environmental Studies in the Chinese Academy of Sciences in Beijing (Ma and Yan, 1989). It was during these early exchanges that we became familiar with Professor Ma's article on ecological engineering that was published in English and in a Western journal (Ma, 1985), describing his definition of ecological engineering as a formal "design with nature."

Professor Ma's associate—Professor Yan Jingsong—from another Chinese Academy of Sciences Institute (Nanjing Institute of Geography and Limnology), who became a great personal friend of our family, proudly described case studies in mid-China designed for improving water quality using ecological

Figure 4.3 American ecological engineering researchers from the Ohio State University and their family members after arrival in Shanghai, China, early May 1989. Left, fish ecologist and Professor Dave Johnson with daughter Amanda Johnson and wife Joyce Johnson; Right, Professor Bill Mitsch with daughter Rebecca Mitsch and wife Ruthmarie Mitsch.

(Photo is property of W.J. Mitsch; taken by unidentified Chinese host for William J. Mitsch.)

engineering principles. One was a case study that used water hyacinths (*Eichhornia crassipes*) to clean a coastal river in an eastern suburb of Suzhou. The water hyacinths decreased the concentrations of phosphorus, nitrogen, COD, and other nutrients in the river and were then harvested and used as feed for fish culture ponds, duck farms, pig farms, and oxen farms. **Figure 4.4** illustrates the water hyacinth treatment system in Suzhou that continued to operate for several years into the 1990s.

4.3 Ecological engineering with salt marsh plantations

One of our early—and most satisfying collaborations in China for me—was with Professor C.H. Chung, a distinguished professor of coastal ecology from Nanjing University (**Figures 4.5** and **4.6**). Above all, he was a first-rate scholar

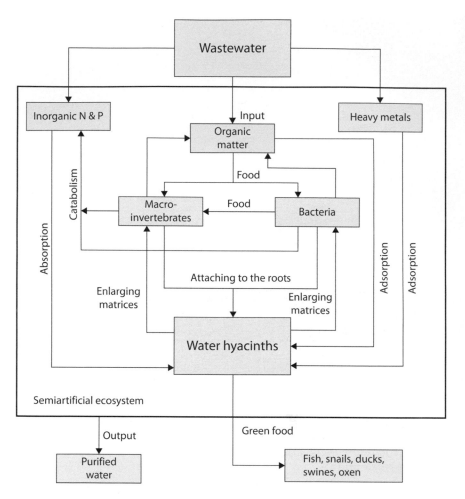

Figure 4.4 *Water hyacinth ecological engineering experiment in Suzhou, China, illustrating the Fumen River that was cleaned for several years by water hyacinth uptake and the advantage to fish, snails, ducks, pigs, and oxen where the hyacinths were used to support green food.*

(From Ma and Yan, 1989. Permission provided by: John Wiley & Sons.)

on coastal ecology and especially on salt marsh restoration and conservation; second, he had received his Ph.D. in the 1930s from The Ohio State University where I was a professor. When we first met, he asked me how the football Buckeyes had done since he left Columbus after his graduation in the 1930s. I knew we would have a fine relationship.

Chung (1989) realized, even 50 years ago, that sea-level rise due to climate change was inevitable and that this sea-level rise would lead to "flooding of coastal plains, aggravation of catastrophic storm tides, intrusion of salt water,

Figure 4.5 *Professor C.H. Chung, Nanjing University, Nanjing, China with experimental Spartina grass in 1989.*

(Photo taken and permission provided by William J. Mitsch.)

and erosion of tideland." He was far ahead of his contemporary coastal scientists. Chung became a *Spartina* (cord grass) scholar and had two basic ecological principles in his research: 1. "The capability of plant communities to modify their environment and 2) the existence of wetland plant community's harsh conditions."

Early in his lab's work, Professor Chung chose *Spartina anglica* as the plant most likely to stabilize China's coastline and be ecologically appropriate as "green manure" for invertebrates and fish. Chung and his lab partners spent decades on small-scale and large-scale studies.

Figure 4.6 *Bill Mitsch visiting Professor C.S. Chung in 2006 after Chung's retirement. Professor Chung is holding material related to his honorary degree at The Ohio State University that was conferred in 1999.*

(Photo property of William J. Mitsch; taken by unidentified Chinese host at Professor Chung's home for William J. Mitsch.)

Chung's studies were highly praised in the beginning of his career. In fact, I nominated him for an honorary degree and it was approved in 1999 at The Ohio State University for his lifetime achievements in stabilizing China's coastline. But when *Spartina alterniflora*, the Eastern USA coastline dominant cord grass, began to invade the West coast of the USA and "took over" unvegetated coastlines, *Spartina alterniflora* became an undesirable species on the U.S. West coast. It was not long before the "invasion" by *Spartina* was critically reviewed by Chinese coastal scientists as well.

4.4 Long-term collaboration with East China Normal University

I have had several long-term relationships with programs and universities near Shanghai partly because of my initial collaboration with Dr. Jianjian Lu (**Figure 4.7**), who was our translator during our first visits to China in 1987 to 1989. So it was natural for us to match up with him again when he secured a professorship at East China Normal University later on.

Figure 4.7 *Professor Jianjian Lu (East China Normal University), Dr. Jingsong Yan (Nanjing CAS Lab on Geography), and Dr. Wenshan He (East China Normal University) (left to right) on a boat cruise on the Pacific Ocean coastline.*

(Photo taken and permission given by William J. Mitsch.)

We had several exchanges of students from Professor Lu's labs to my Ohio and Florida labs. For example, Ms. Bing-Bing Jiang completed her Ph.D. under me at University of South Florida in 2015–2020 and then held a post-doc position in our lab at FGCU from 2020 to 2021. Her Ph.D. thesis was titled "Exploring the potential of nutrient retention and recycling with wetlaculture systems in Ohio with physical and landscape models." The dissertation was mostly based on a new concept of ours called "wetlaculture"—a novel way for solving agricultural nutrient runoff with wetlands. This approach dominated my last decade as a professor and thesis advisor; we were able to publish a summary of wetlaculture with many co-authors recently (Mitsch et al., 2023).

Dr. Jiang returned to China in 2021 after she completed her degree and post-doc and was immediately offered a faculty position in The School of Fishery, Zhejiang Ocean University, Dinghai District, Zhoushan, in China, where she continues to teach ecological engineering. Two papers (Jiang et al., 2021; Jiang and Mitsch, 2020) and several conference abstracts were published as a direct result of her six years in our laboratories.

We also collaborated with Professor Lu and Dr. He from Lu's lab on a key paper providing an update on the ecological condition on the pool behind the Three Gorges Dam on the Yangtze River (Mitsch et al., 2008). This paper is one of the more significant in my career because it was a collaboration between the West and China that was published in *Science* and because of the importance of the Three Gorges Dam on dam management (it is one of the largest man-made reservoirs and hydroperiods in the world). It illustrates an ambitious approach that China is using to adapt to both arid and excessive water climates, often at the same time.

4.5 Investigating the ecological balance of Chinese fishponds

The agriculture-aquaculture systems in China have two main parts: fishponds and cropped dikes (Yan and Yao, 1989). The key concept in these ponds and landscapes is recycling. Human and livestock wastes are also thrown into the ponds for nutrition, and pond's bottom mud, after the growing season, is applied to the dikes as fertilizer.

In the early 1990s, my lab at The Ohio State University decided to investigate, with dynamic modelling, if the recycling really did contribute to better productivity in these ponds. The computer model was developed using STELLA (Systems Thinking, Experiential Learning Laboratory with Animation) II version 1.0.2 on our Macintosh computers. Solar radiation was randomly generated with monthly averages based on in situ data. All other processes were assumed to be deterministic, expressed in differential equations with respect to time. A fourth-order Runge-Kutta integration method with a one-day time step was adopted as a base run. Simulation was carried out for a growing season from April 1 until November 15 for a total period of 230 days. Details of the model are found in Hagiwara and Mitsch (1994).

Our modeling process followed the procedure suggested by our friend Sven Erik Jørgensen and my major professor H.T. Odum—a dynamic, deterministic model that assumes the system consisting of energy flows, represented by differential equations with respect to time, between compartments. The selection of the state variables shown in **Figure 4.8** was based on the separation of distinct niches occupied by the organisms in the ecosystem. To assess the relative contribution of key parameters to the outcome of the model, sensitivity analysis was performed. Six fish species were included in the model: Common Carp (*Cyprinus carpio*), Wuchang Fish (*Megalobrama amblycephala*), Bighead Carp (*Aristichthys nobilis*), Silver Carp (*Hypophthalmichtys molitrix*), Grass Carp (*Ctenopharyngodon Idella*), and Black Carp (*Mylopharyngodon piceus)* (**Figure 4.8**).

One of the important findings of our model simulations was the illustration of the importance of species diversity on the productivity of other animals, plants, microbes, and detritus in the fishpond (**Table 4.2**; Hagiwara and

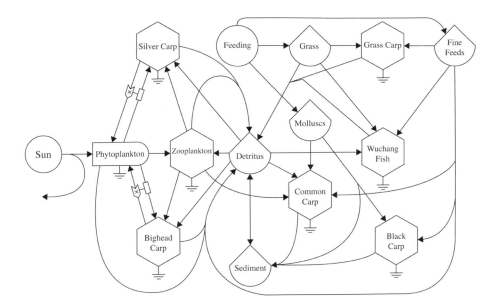

Figure 4.8 *Odum diagram of our fishpond food chain model.*

(From Hagiwara and Mitsch, (1994). Permission provided by Elsevier Science.)

Mitsch, 1994). For example, the role of detritus as a food for Silver Carp and Wuchang Fish suggests that these fish can adapt as detritus becomes rare. As summarized by Hagiwara and Mitsch (1994): "For Common Carp, its removal resulted in higher growth for competing species..." The balance in the modeled fishpond and in the many study ponds in China is delicate because there are so many feedbacks in these multi-species systems. The lessons from these fishponds, based on centuries of empirical evidence, are important for aquatic ecologists to note: Diversity does matter and can affect productivity of ecosystems, sometimes positively and sometimes negatively.

Table 4.2 Effect on Chinese Fishpond Ecosystem as Shown in Figure 4.8 of Removal of Species

Species	Effect of Removal
Silver Carp	Lower DO; higher zooplankton; larger fluctuations of primary production
Bighead Carp	Persistence of zooplankton; virtual absence of phytoplankton
Grass Carp	Less detritus; higher DO; less phosphorus but little overall change
Wuchang Fish	Least overall change except better growth for Black Carp and Common Carp
Black Carp	No Silver Carp growth due to decreased detritus
Common Carp	Higher fish growth except for Silver Carp and Wuchang Fish due to less competition; higher zooplankton biomass and lower amount of detritus

Source: Revised from Hagiwara and Mitsch, 1994
Abbreviation: DO, Dissolved oxygen

4.6 Recent collaborations with Chinese ecology and wetland programs

4.6.1 Northeast Normal University, Changchun, China

Later in our exchanges with China, we became aware of significant wetland science and management practices and opportunities for research based at Northeast Normal University in the northeast "neck of China" centered on Changchun (**Figures 4.9** and **4.10**). There are notable wetland centers and labs in the region and we were welcomed there several times in our last decade of China visits. The wetlands in this part of China were dominated by many species of cranes, such as red-crowned cranes that attracted many ecotourists to this region of China.

4.6.2 Other enjoyable ventures to Chinese meetings and labs

China began to host significant international wetland meetings at the end of the 20th century and continuing in the 21st century, including the 10th INTECOL Wetland Conference in Changshu, China during September 2016 (**Figure 4.11**); Coastal Blue Carbon and Wetland Restoration Conference, Wenzhou, Zhejiang, China, in August 2017; and the Society of Wetland Scientist Asian Chapter meeting in China. Other memorable visits to China occurred

Figure 4.9 *Bill Mitsch with Lianxi Sheng, Director, Key Laboratory of Wetland Ecology, Northeast Normal University, Changchun, China.*

(Photo provided by Northeast Normal University, Changchun, China, to Professor W.J. Mitsch.)

Figure 4.10 *Wetland Conference Attendees, Northeast Normal University, Changchun, China.*

(Photo provided by Northeast Normal University, Changchun, China, to personal collection of Professor W.J. Mitsch.)

Figure 4.11 *Participants at 10th INTECOL Wetlands Conference "Healthy Wetlands, Healthy Earth," Changshu, China, September 2016.*

(Photo property of W.J. Mitsch, taken by unidentified Chinese host.)

Figure 4.12 *Xuehua Liu, left, Professor of School of Environment, and her students at Tsinghua University, Beijing, China.*

(Photo property of W.J. Mitsch taken by unidentified Chinese host.)

when I was invited to a very fine university in Beijing, Tsinghua University, by Professor Xuehua Liu (**Figure 4.12**).

I simply enjoyed it when I gave lectures in classrooms or auditoriums with little fanfare; I only needed a microphone, a power point projector, and an audience (**Figure 4.13**).

Since it is unlikely that I will travel to China much more now because I am formally retired and getting old, I have already substituted travel around the world that takes enormous effort and cost with giving remote lectures to wetland and water resources programs and departments. For example, I was invited to give a short remote presentation in late October 2022 to friends and their students and staff at Northeast Normal University in Changchun. The Chinese cut and pasted my presentation to be a keynote presentation that was seen on November 19–21, 2022 at a remote meeting titled:

The First Wetland Environmental Ecology Summit Forum

Wetland Environmental Ecology and Global Change

More than 1,500 remote attendees participated in this workshop.

Figure 4.13 *Bill Mitsch lecturing in Nanjing, China, in May 2017.*

(Photo property of W.J. Mitsch taken by unidentified Chinese host.)

References

Chung, C-H. 1989. Ecological engineering of coastlines with salt marsh plantations. In: Mitsch, W.J. and S.E. Jørgensen, eds., *Ecological Engineering: An Introduction to Ecotechnology*. John Wiley & Sons, Inc., New York, NY. 255–289 pp.

Costanza, R. (ed.). 1991. Ecological Economics: The Science and Management of Sustainability. Columbia University Press, New York, NY.

Hagiwara, H. and W.J. Mitsch. 1994. Ecosystem modeling of a multi-species integrated aquaculture pond in South China. *Ecological Modelling* 72: 41–73.

Jiang, B.B. and W.J. Mitsch. 2020. Influence of hydrologic conditions on nutrient retention and soil and plant development in a former central Ohio swamp: A wetlaculture mesocosm expriment. *Ecological Engineering* 157: 105969.

Jiang, B.B., W.J. Mitsch and C. Lenhart. 2021. Estimating the importance of hydrologic conditions on nutrient retention and plant richness in a wetlaculture mesocosm experiment in a former Lake Erie basin swamp. *Water* 13: 2509.

Ma, S. 1985. Ecological engineering: Application of ecosystem principles. *Environmental Conservation* 12(4): 331–335.

Ma, S. and J. Yan. 1989. Ecological engineering for rreatment and utilization of wastewater. In: Mitsch, W.J. and S.E. Jørgensen, eds., *Ecological Engineering: An Introduction to Ecotechnology*. John Wiley & Sons, Inc., New York, NY. 185–217 pp.

Mitsch, W.J. 1987. Ecological engineering: The roots and rationale of a new ecological paradigm. In: C. Etnier and B. Guterstam (eds.), *Ecological Engineering for Wastewater Treatment*. CRC/Lewis Publishers, Boca Raton, FL. 1–20 pp.

Mitsch, W.J. 1991. Ecological Engineering—approaches to sustainability and biodiversity in U.S. and China. In: R. Costanza (ed.), *Ecological Economics: The Science and Management of Sustainability*. Columbia University Press, New York, NY. 428–448 pp.

Mitsch, W.J. and S.E. Jørgensen (eds.) 1989. *Ecological Engineering: An Introduction to Ecotechnology*. John Wiley & Sons, New York, NY. 472 pp.

Mitsch, W.J., J. Lu, X. Yuan, W. He and L. Zhang. 2008. Optimizing ecosystem services in China. *Science* 322: 528.

Mitsch, W.J., B. B. Jiang, S. Miller, K. Boutin, L. Zhang, A. Wilson and B. Bakshi. 2023. Wetlaculture: Solving harmful algal blooms with a sustainable wetland/agricultural landscape. In: B. R. Bakshi (ed.), *Engineering and Ecosystems: Seeking Synergies for a Nature-Positive World*. Springer, New York, NY.

Yan, J. and H. Yao. 1989. Integrated fish culture management in China. In: Mitsch, W.J. and S.E. Jørgensen (eds.), *Ecological Engineering: An Introduction to Ecotechnology*. John Wiley & Sons, Inc., New York, NY. 375–408 pp.

Using Wetland Science to Support Federal Protection of Wetlands as "Waters of the United States"

Students doing wetland delineation.

(Permission provided by William J. Mitsch.)

DOI: 10.1201/9781003374619-5

5.1 Introduction

The protection of wetlands by the federal government in the USA has been a half-century struggle since the Feds were brought into wetland protection not by Congressional mandate, but because of a lower court interpretation of a Congressional law, also known as the Federal Water Pollution Control Act (PL 92-500), more frequently described as the Clean Water Act as amended in 1972, 1977, and 1982. Most of the Clean Water Act in the 1970s was assigned to the then-new U.S. Environmental Protection Agency (EPA), but Section 404 of the Act, administrating a dredge-and-fill permit program, had been the responsibility of the U.S. Army Corps of Engineers for the entire 20th century, so it was reassigned to the Corps in the Clean Water Act. Initially, in Section 404, the Corps was asked to grant Section 404 permits to only "navigable waters." But the definition of waters of the United States (WOTUS) was expanded to include wetlands in two court decisions in 1974–75—*United States v. Holland* and *Natural Resources Defense Council v. Calloway*—and these decisions then put the Army Corps of Engineers squarely in the center of wetland protection in the United States. On July 25, 1975, the Corps issued revised regulations for the Section 404 program that enunciated the policy of the United States on wetlands:

> As environmentally vital areas, [wetlands] constitute a productive and valuable public resource, the unnecessary alteration or destruction of which should be discouraged as contrary to the public interest.
>
> **—Federal Register, July 25, 1975**

Wetlands were not mentioned in the early versions of the Clean Water Act and the U.S. Army Corps of Engineers were not familiar with what wetlands were, let alone how to define them. By these court decisions, the jurisdiction of the Corps was extended to include 150 million acres (60 million ha) of wetlands, 45% of which are found in Alaska. Even a decade later, in 1985, the U.S. Supreme Court, in *United States v. Riverside Bayview Homes*, rejected the contention that Congress did not intend to include wetland protection as part of the Clean Water Act (Mitsch and Gosselink, 2015).

The procedure for obtaining a "404 permit" for dredge-and-fill activity in wetlands is complex. In the initial screening of a project that involves potential effects on wetlands, the following three approaches are evaluated in sequence:

1. *Avoidance*: Taking steps to avoid wetland impacts where practicable.
2. *Minimization*: Minimizing potential impacts to wetlands.
3. *Mitigation*: Providing compensation for any remaining, unavoidable impacts through the restoration or creation of wetlands, and more recently by paying in lieu fees.

5.2 Wetland delineation and delineation manuals

To determine whether a particular piece of land was a wetland and therefore if it was necessary to obtain a Section 404 permit to dredge or fill that wetland, federal agencies, beginning with the Army Corps of Engineers, developed guidelines for the demarcation of wetland boundaries in a process that came to be called *wetland delineation*.

As soon as they were given the responsibility for protecting wetlands under the Clean Water Act, the U.S. Army Corps of Engineers snapped to attention, saluted, and published a technical manual for wetland delineation (*1987 Corps of Engineers Wetlands Delineation Manual*), the first of three wetland delineation manuals proposed by the federal government in a five-year period (**Figure 5.1a**). This first manual specified three mandatory technical criteria—relating to hydrology, soils, and vegetation—for a parcel of land to be declared a wetland. This manual was immediately accepted by consultants as the way to define wetlands.

Subsequently, the U.S. EPA, the Soil Conservation Service, and the U.S. Fish and Wildlife Service joined the U.S. Army Corps of Engineers and began to

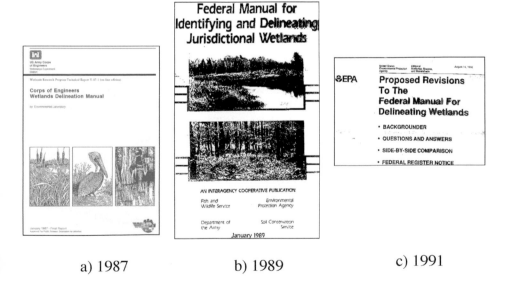

a) 1987   b) 1989   c) 1991

Figure 5.1 *Three wetland delineation manuals were developed in the U.S in (a) 1987, (b) 1989, and (c) 1991 by different federal agencies. After several years of heated discussion, the USA returned to the first manual from 1987 as written by the U.S. Army Corps of Engineers; that wetland delineation manual is still used today.*

(Federal property not subject to copyright. Photo permission by William J. Mitsch.)

develop a separate interagency document that involved all the main federal agencies. To unify the government's approach to wetlands, months of political and scientific debate and negotiations took place among the agencies, resulting in a *Federal Manual for Identifying and Delineating Jurisdictional Wetlands* (*1989 Wetlands Delineation Manual;* **Figure 5.1b**), published in August 1989. This 1989 manual, while also requiring the three mandatory technical criteria for a parcel of land to be declared a wetland, allowed one criterion to infer the presence of another (e.g., the presence of hydric soils to infer hydrology). The manual also provided some guidance about how to use field indicators such as watermarks on trees or stains on leaves to determine recent flooding, wetland vegetation (from published lists), and hydric soil indicators such as mottling.

The second wetland delineation manual led to a contentious and quite vocal period in U.S. wetland policy history, where the second delineation manual was considered by some to be too liberal in defining wetlands. Furthermore, it had never been published as a draft for public comment.

That situation led to the development of a third wetland delineation manual from the Executive Branch of the government (the White House) through the U.S. EPA, *Proposed Revisions to the Federal Manual for Delineating Wetlands* (*1991 Wetland Delineation Manual;* **Figure 5.1c**). The draft 1991 manual was prepared in response to heavy lobbying by developers, agriculturalists, and industrialists for a relaxation of the 1989 wetland definitions to lessen the regulatory burden upon the private sector. That manual was published for public comment in August 1991 but was quickly and heavily criticized for its lack of scientific credibility and its unworkability (Environmental Defense Fund and World Wildlife Fund, 1992). This third wetland delineation draft called for extensive field testing by four federal panels, as well as one independent panel. The dispute passed through my office, beginning in August 1991, when I received a phone call from the U.S. EPA and was asked if I would be willing to chair an "independent" panel, with committee members chosen by me, to define wetlands, parallel with other federal review panels. Thinking it would be a simple exercise to organize a committee to test this new manual, I agreed to chair the independent panel. In the meantime, wetlands were beginning to be discussed on the front and editorial pages of major newspapers almost every day and new political cartoons about wetlands also appeared quite frequently, because this new proposed wetland delineation manual employed restrictive wetland definitions developed by the George H. W. Bush administration, specifically by Vice President Dan Quayle's Council on Competitiveness, based in the White House.

Chairing the "independent" panel proved to be a disaster because no matter whom I proposed to be on my panel, I was told that "the White House" or some unnamed administrator in the White House vetoed my nominee or nominees for the committee. This exercise became a little ridiculous after two months of my providing nominees for my panel and being told that the they were unacceptable, supposedly because they had already made a public statement opposing the new wetland rules. Finally, I had no recourse other than to resign as chair of the committee. And that made headlines in the U.S. EPA newsletter, then and still called *INSIDE EPA*, with a heading noting that a scientist (Mitsch) had resigned the wetland panel amid controversy.

That story in *INSIDE EPA* must have caught the attention of legislators and congressional committees that were already publicly discussing the new wetland definitions and laws. For example, a fellow wetland scientist, Professor Margorie M. Holland from the University of Mississippi, was attending one such congressional committee meeting discussing wetlands in the fall 1991 in Washington, DC, on behalf of the Ecological Society of America, which had already written an opinion piece in opposition to the new August 1991 wetland delineation rules. She contacted me during or immediately after the committee deliberations and asked me, "Bill, were your ears burning?" The Congressional Committee contacted me shortly after, wanting me to come and testify in Washington, DC. I recall that I did send the committee a written statement telling them that I was unable to secure any committee members who were acceptable to the White House.

Marge Holland's Recollection

Just prior to the 1991 annual meeting of the Ecological Society of America (ESA), the ESA staff learned of an announcement in the *Federal Register* about proposed changes to the national wetland delineation criteria. Five ESA members were charged with conducting a scientific review of the proposed revisions to the 1989 *Federal Manual for Identifying and Delineating Jurisdictional Wetlands*. Several ESA members presented testimony to three Congressional Science Committees during the Fall and Winter of 1991–92 that included the statement that the 1991 wetlands revisions were "fundamentally flawed." Their testimony included examples from the *Wetlands* book, co-authored by William Mitsch and James Gosselink, to explain why the 1991 proposed revisions were flawed. When Holland and Mitsch contacted each other later, Holland asked Mitsch, "Bill, were your ears burning?" noting the frequent use of material from his book during ESA members' testimonies.

5.3 Vice President Dan Quayle's White House Council on Competitiveness

I found out later that, unfortunately, everybody I had proposed to the U.S. EPA for my committee over several months was rejected by Vice President Quayle's "White House Council on Competitiveness." Watzman and Triano (1992) described this council as employing "shadowy activities" and stated that

> the Council had steadily built its reputation by working behind the scenes to undermine health, safety, and environmental regulations... from the Clean Air Act to nutrition labeling.... By acting as a super-agency, with the power to review all regulations, the Quayle council not only adds another layer to an already lengthy process, but it also defies the basic principles upon which the regulatory edifice is built. When a federal agency writes rules, it is required by law to hear from all sides, make decisions only on the merits and make communications available at a public docket where anybody can look at them. The Quayle council does not follow any of these open-government standards.

After I resigned from the position as the "independent panel chair," I publicly supported a campaign that was put forward by others insisting that these vital decisions on how a wetland is defined needed input from the most prestigious science address in Washington DC—The National Academy of Sciences (NAS). I was invited to Washington, DC, to give a presentation to the NAS soon after my panel chair resignation. I described to the NAS what had happened with my panel and encouraged them to become involved. I recall that one NAS member came up to me after my presentation and calmly suggested that the NAS commonly takes on difficult scientific issues, so this wetland definition question should not be much of a problem. I did not dispute his confidence then, but I did know even then that the "Academy" would be in for a big fight on the wetlands issue.

Ultimately, the NAS was invited by the U.S. EPA to organize a panel that pretty much would do what my independent panel would have done several years earlier. But the NAS panel would not get started until 1993, as it took two years for the U.S. EPA to find funds to pay for the NAS-led review. The NAS committee had 17 wetland, hydrology, and soil experts. I was one of those 17. In the meantime, the nation also had elected a new president—Bill Clinton—and he was much more wetland-friendly.

5.4 New wetland-threatening bills from Congress in 1995

Early in 1995, with wetland opponents seeing that redefining wetlands was not going to work, House and Senate bills were being introduced by Congress that would drastically change wetland management in the USA. The National Research Council (NRC) leaders brought this effort to our attention and asked us to speed up the writing of our report. These proposed bills—House Bill 961 and Senate Bill 851—would have protected only 15 million hectares (37 million acres) of wetlands out of the remaining 42 million hectares (102 million acres) of wetlands in the United States (**Figure 5.2**).

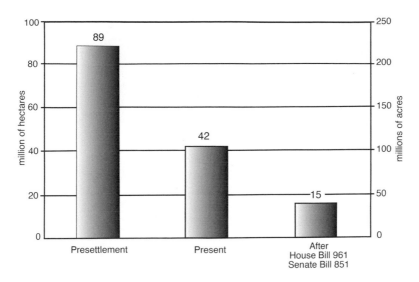

Figure 5.2 *Estimated extent of wetlands in the lower 48 states of the United States for presettlement times (1780s) and 1990s. The numbers in the first two bars are compared with an estimate of the extent of wetlands that would have remained legally protected if House Bill 961 or Senate Bill 851 in the U.S. Congress had been passed in 1995.*

(From Mitsch and Gosselink, 2015. Permission from John Wiley & Sons, Inc.)

We would have lost protection of 27 million hectares (65 million acres) in the lower 48 states if these bills had passed. We scientists recognized that the members of Congress were not trained to understand how wetlands functioned any more than we scientists could write a Congressional bill. For example, the House bill required that land be flooded to the surface for 21 consecutive days in the growing season, a ridiculously wet expectation for many wetlands (Cushman, 1995). Carol Browner, the director of the U.S. EPA, summarized "that wetland protection and the definition of wetlands, has to be based on sound science, not on political expediency" (Cushman, 1995).

In the end, partially because of the release of our report in the nick of time, neither bill passed; but this close call illustrated that wetlands can be lost either by drainage or by any legal fiat that redefines wetlands based upon poor science. On May 10, 1995, the NAS released our report (Lewis et al., 1995) on wetland characterization during a House debate on the new bill. Both House and Senate bills went nowhere, and our nation's wetlands were saved (**Figure 5.3**). The headline in the Wednesday May 10, 1995, New York Times was "Scientists Reject Criteria for Wetlands Bill." We were a bunch of proud wetland scientists who used the best science to define what wetlands were and probably saved a lot of wetlands in the USA.

Figure 5.3 *The hardcover version of the National Research Council/National Academy of Sciences report "Wetlands: Characteristics and Boundaries" (Lewis et al., 1995) as finally published in hard copy in summer 1995.*

(Permission from National Research Council/National Academy of Sciences, Washington DC.)

In the meantime, the 1987 Corps technical manual (U.S. Army Corps of Engineers, 1987), generally agreed to be a version ecologically and politically between the "liberal" 1989 manual and the "conservative" 1991 manual, continued to offer the unofficial test by which wetlands are determined, and it continues to be used to today. This five-year (1991–1995) fight was worth every minute invested in it in the early 1990s. Nevertheless, the threat by Trump's administration 25 years later showed that we cannot rest on our laurels (Davenport, 2020).

5.5 Update on the 2020 Trump water rule

In June 2021, President Joe Biden's administration announced that they are planning on rolling back the 2020 Trump rule that would have allowed unregulated pollution of streams and rivers and in particular would have removed many wetlands from the Waters of the United States protection (Silverberg, 2021). Both the U.S. EPA, through their new administrator Michael Regan, and the U.S. Army Corps of Engineers, were clear in their statements and press releases that the Trump definitions were unacceptable to their respective agencies. Wetlands may have been saved from the brink yet again, but the Supreme Court is once again determining whether they will consider the definition of wetlands for the fourth time in the 21st century (Raymond and Chung, 2022, Vock, 2022).

References

Cushman, J.H. 1995. Scientists reject criteria for wetlands bill. *New York Times*, May 10, 1995, Section D, page 19.

Davenport, C. 2020. Trump removes pollution controls on streams and wetlands. *New York Times*, January 22, 2020. https://www.nytimes.com/2020/01/22/climate/trump-environment-water.html

Environmental Defense Fund and World Wildlife Fund. 1992. How Wet Is a Wetland? The Impact of the Proposed Revisions to the Federal Wetlands Delineation Manual. Environmental Defense Fund and World Wildlife Fund, Washington, DC. 175 pp.

Lewis, W.M., B. Bedford, F. Bosselman, M. Brinson, P. Garrett, C. Hunt, C. Johnston, D. Kane, A.N. Macrander, J. McCulley, W.J. Mitsch, W. Patrick, R. Post, D. Siegel, R.W. Skaggs, M. Strand, and J.B. Zedler. 1995. *Wetlands: Characteristics and Boundaries*. The National Academies of Science Press, Washington, DC.

Mitsch, W.J. and J.G. Gosselink. 2015. *Wetlands*, 5th ed. John Wiley & Sons, Inc., Hoboken, NJ. 744 pp.

Raymond, N. and A. Chung. Oct. 2022. U.S. Supreme Court leans toward limiting wetlands regulation. https://www.reuters.com/legal/us-supreme-court-leans-toward-limiting-wetlands-regulation-2022-10-03/

Silverberg, D. 2020. FGCU wetlands professor blasts Trump water rules, calls for citizen action. *The Paradise Progressive* blog. https://theparadiseprogressive.home.blog/2020/01/24/fgcu-wetlands-professor-blasts-trump-water-rules-calls-for-citizen-action

Silverberg, D. 2021. FGCU wetlands expert Bill Mitsch hails Biden rollback of Trump water rule. *The Paradise Progressive* blog. https://theparadiseprogressive.home.blog/2021/06/14/fgcu-wetlands-expert-bill-mitsch-hails-biden-rollback-of-trump-water-rule/

U.S. Army Corps of Engineers. 1987. Corps of Engineers Wetland Delineation Manual. Technical Report Y-87-1. U.S. Army Corps of Engineers Waterways Experiment Station, Vicksburg, MS, 100 pp and appendices.

Vock, D.C. 2022. Supreme Court could shift more control over wetlands to states. Route Fifty. Washington DC, https://www.facebook.com/routefifty/

Watzman, N. and C. Triano. 1992. Defund Quayle's autocratic competitiveness council regulatory agencies: His secretive group blocks health, safety, and environmental policies at the behest of business. *Los Angeles Times*, June 24, 1992. https://www.latimes.com/archives/la-xpm-1992-06-24-me-697-story.html

6

Reconnecting Rivers to Their Floodplains

A bald cypress swamp, known locally as Heron Pond, in southern Illinois, at the Cache River State Natural Area—one of my study sites while I was at IIT in the 1970s.

DOI: 10.1201/9781003374619-6

6.1 Introduction

Rivers and their floodplains are important hydrologic connections on our landscapes (Mitsch et al., 1979). For most of recorded history in developed regions, society has separated floodplains from their rivers, leading to landscapes that no longer fit their hydrologic conditions. Because floodplains flood typically two years out of three, at least in our Midwestern landscapes, and not permanently, societies found that it was relatively easy with ditches and dikes to "unflood" floodplains permanently and create more land for human living and agriculture. The benefits were supporting higher human density per unit of land area.

We often see that homeless people camp on floodplains because, to them, the floodplains are usually dry and only occasionally wet. But most people do not like living where it floods even occasionally. I have spent a good percentage of my career investigating how rivers are connected to their floodplains and how water quality and quantity are impacted when those connections are disrupted.

This chapter presents two studies, both early in my career and both in Illinois, of natural and enhanced floodplain-river exchanges in the Midwest. The first site was river flooding that was still natural, robust, and somewhat predictable; the second site was a partially drained floodplain where scientists were investigating what happens when flooding to drained floodplains is reintroduced to the biology and biogeochemistry of the river and its floodplain.

6.2 Southern Illinois floodplain wetlands

When I returned to Illinois from Florida in 1975 for my first faculty position at Illinois Institute of Technology (IIT) in Chicago after my graduate studies at University of Florida, I lamented that I had left my favorite ecosystems in Florida—cypress-tupelo swamps. These riverine swamps were sometimes referred to as cathedrals or sanctuaries because of their majestic appearance. I missed my swamps! However, a knowledgeable natural resource scientist from the University of Illinois told me, to my astonishment, that Illinois indeed had many cypress (*Taxodium*) swamps, and they were mostly found in the southern tip of Illinois, where the great rivers of North America—the Ohio, the Missouri, and the Mississippi—merged. I took the bait and visited the floodplains of southern Illinois, especially the Cache River, whose history was connected to these mighty rivers. I was directed to Heron Pond, one of the most majestic backwater riverine floodplain cypress swamps I have ever seen anywhere, tucked away into the northern reaches of the Mississippi River embayment (**Figure 6.1**). It reminded me of the cypress swamp slow-flowing sloughs that have such great biodiversity. I had a few good grad students,

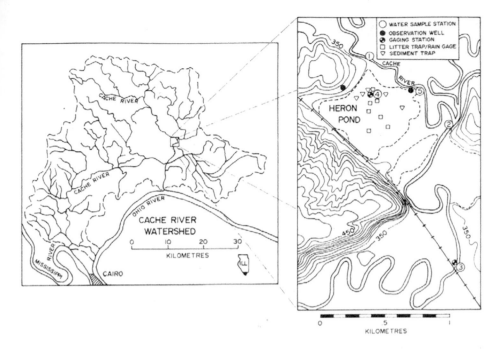

Figure 6.1 *Cache River watershed in southern Illinois, including water quality sampling locations and sediment traps in and around Heron Pond.*

(From Mitsch et al., 1979.)

led by Carol Dorge and John Wiemhoff, who helped me describe the hydrologic and nutrient connections between the flooding Cache River and the flooded Heron Pond. We were fortunate to be able to gather water quality and sedimentation data during an early spring flood in March 1977 (**Figure 6.2**) and were able to conduct a three-year study at Heron Pond in southern Illinois while I was a professor at IIT. "The perfect storm" of a river flooding in the middle of the study that March allowed us to develop a complete phosphorus budget describing an important biogeochemical connection between the river and its floodplain (Mitsch et al., 1979; **Figure 6.3**) that we published in the high-impact journal *Ecology*.

We were able to quantify that this river and backwater swamp exchange was first and foremost the way that floodplains are naturally fertilized by rivers. We found that sedimentation by the river deposited just a tiny percentage of the river's phosphorus onto the floodplain that was probably enough "fertilizer" to support massive-sized cypress and tupelo trees and the ecosystem structure for many organisms. We also found that one flood brought in ten times more phosphorus than what flowed back to the river for the rest of the year. For that year, the cypress swamp was a nutrient sink, essentially cleaning up the river of excessive phosphorus.

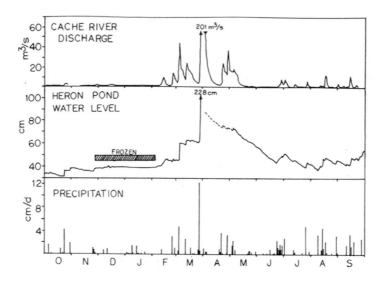

Figure 6.2 *Hydrologic conditions of floodplain Heron Pond cypress swamp and Cache River that seasonally floods the swamp. These data from 1977 to 1978 are from the site described in* **Figure 6.1.**

(From Mitsch et al., 1979.)

6.3 The Des Plaines River Wetland Demonstration Project

Several years later, just as I was moving to Ohio State in late 1985, I had the opportunity to collaborate with Dr. Donald L. Hey (RIP July 24, 2022) and other wetland scientists involved with his innovative research at the Des Plaines River Wetland Demonstration Project in Lake County, Illinois, where he was using pumps to optimize exchange between the Des Plaines River and created riparian wetlands. Don's favorite animal was the beaver because of its ability to build dams on rivers and essentially change the flux of water and nutrient exchanges between rivers and their floodplains, just as we were doing with the artificial floodplain marshes on the Des Plaines River. Again, I benefited from the right mixture of grants, grad students, wonderful collaborators, and being in the right place at the right time. A flood of publications came out of my lab at The Ohio State University on hydrology, water quality vegetation, and aquatic organisms at the site during this study period, with data reported in Cronk and Mitsch (1994a,b); Fennessy et al. (1994); Sanville and Mitsch (1994); Mitsch et al. (1995); and Wang and Mitsch (2000).

Located 60 km north of Chicago, Illinois, USA, The Des Plaines River Wetlands Demonstration Project includes a 4.5 km (2.8 mile) stretch of the Des Plaines River and more than 180 hectares (445 acres) of prairie, forest, and passive and experimental wetlands. The Des Plaines River drains a watershed of ~545 km² (210 mi²) (80% agricultural and 20% urban), and its waters are polluted with

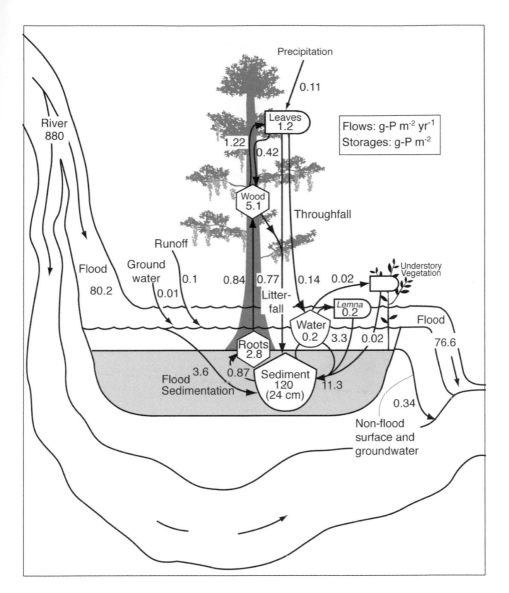

Figure 6.3 *A phosphorus budget of a backwater cypress-tupelo swamp named Heron Pond on the floodplain of the Cache River in southern Illinois. (Diagram redrawn from Mitsch et al., 1979; and published in Mitsch and Gosselink, 2015.)*

(Permission provided by John Wiley & Sons, Inc.)

sediment and nutrient concentrations typical of the midwestern USA. Within the study site, four basins, numbered 3, 4, 5, and 6, were between 1.9 and 3.4 ha (4.7 and 8.4 acres) in size and were constructed in 1986–1988 (**Figure 6.4**). In 1989, our river pumping experiment began and the basins were flooded with Des Plaines River water each day. The experimental wetlands were

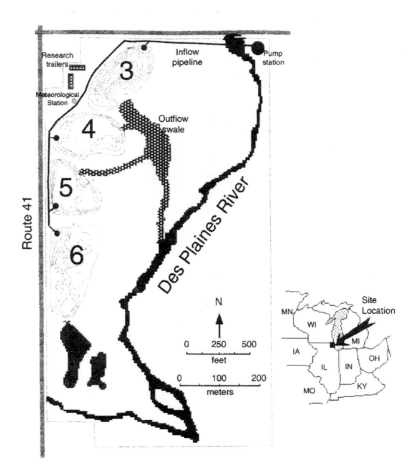

Figure 6.4 *The Des Plains River Wetland Demonstration Project in Lake County, Illinois, illustrating the original four experimental basins in a three-year hydrologic loading experiment. (Figure from Wang and Mitsch, 2000.)*

(Permission provided by Elsevier.)

hydrologically isolated from one another; pumped river water enters each wetland at one end and exits through one outflow point into swales leading back to the river. Pumping rates to the wetland basins were maintained to test the influence of two hydrologic flowthrough conditions: two experimental wetlands (EW) were subjected to high-flow conditions (EW 3 and 5; design flows of 30 cm/week; 12 inches/week) and two to low-flow conditions (EW 4 and 6; design flows of 5 cm/week; 2 inches/week). For comparison, average precipitation input during the study was 1.5–1.6 cm/week (0.59–0.63 inches/wk), and little if any lateral runoff entered the wetland other than the pumped water.

All the marshes were dominated by *Typha latifolia* and *T. angustifolia*, although there were anywhere from 18 to 45 wetland vegetation species reported from 1990 to 1992, the period of this study (Fennessy et al., 1994). Some of the other common species in these marshes included *Phalaris arundinacea*, *Polygonum* spp., *Alisma plantago-aquatica*, *Chara* sp., *Eleocharis calva*, *Ludwigia palustris*, *Potamogeton pectinatus*, *Sagittaria latifolia*, and *Scirpus validus*. The only plant that was introduced to these wetlands and survived more than one growing season was native water lily (*Nymphaea odorata*), which was introduced to two of the wetlands in 1989 and flourished in the deepwater areas of EW 3 and 4. Studies of hydrology, water quality vegetation, and aquatic organisms at the site during this study period are reported in Cronk and Mitsch (1994a,b); Fennessy et al. (1994); Sanville and Mitsch (1994); Mitsch et al. (1995); and Wang and Mitsch (2000). We found that the high-flow floodplain wetlands were generally the most productive.

The table was now set for what would be yet another achievement of my career and our studies investigating river-floodplain exchanges—developing a series of floodplain wetlands on the Olentangy River on the campus of The Ohio State University to refine wetland and floodplain science as has never been done before (see Chapter 7). The Olentangy River Wetlands also allowed us to demonstrate how creating wetlands on a college campus broadened our teaching of wetland ecology to thousands of students rather than dozens every year. We found at the Des Plaines River and Olentangy River wetlands that nature was now teaching the students too just as we noted during the Boatload of Knowledge trip on the Ohio River did with graduate students 20 years earlier (see Chapter 2).

References

Cronk, J.K. and W.J. Mitsch. 1994a. Periphyton productivity on artificial and natural surfaces in four constructed freshwater wetlands under different hydrologic regimes. *Aquatic Botany* 48: 325–341.

Cronk, J.K. and W.J. Mitsch. 1994b. Aquatic metabolism in four newly constructed freshwater wetlands with different hydrologic inputs. *Ecological Engineering* 3: 449–468.

Fennessy, M.S., J.K. Cronk, and W.J. Mitsch. 1994. Macrophyte productivity and community development in created freshwater wetlands under hydrologic conditions. *Ecological Engineering* 3: 469–484.

Mitsch, W.J., J.K. Cronk, X. Wu, R.W. Nairn, and D.L. Hey. 1995. Phosphorus retention in constructed freshwater riparian marshes. *Ecological Applications* 5: 830–845.

Mitsch, W.J., C.L. Dorge, and J.W. Wiemhoff. 1979. Ecosystem dynamics and a phosphorus budget of an alluvial cypress swamp in southern Illinois. *Ecology* 60: 1116–1124.

Sanville, W. and W.J. Mitsch (eds.). 1994. Creating freshwater marshes in a riparian landscape: Research at the Des Plaines River Wetlands Demonstration Project. *Special Issue of Ecological Engineering* 3: 315–321.

Wang, N. and W.J. Mitsch. 2000. A detailed ecosystem model of phosphorus dynamics in created riparian wetlands. *Ecological Modelling* 126: 101–130.

7

Battling for a Campus Where Students Can Soak Up Knowledge

The Olentangy River Wetland Research Park and the adjacent Olentangy River at The Ohio State University in 2013. Note the block "O" on the roof of the Sandefur Wetland Pavilion, completed in 1999, (lower left) signifying The Ohio State University to low-altitude aircraft. The white-roofed building on the lower right is the Heffner Wetland Research and Education Building, completed in 2003.

(Photo courtesy of Ohio Department of Transportation.)

DOI: 10.1201/9781003374619-7

December 2, 2011

The Ohio State University, Columbus, USA, announced at 10:56 am this morning December 2, 2011, that the University is turning down the request by Columbia Gas for a natural gas pipeline easement under the Olentangy River Wetland Research Park on campus. The ORWRP was the 24th Ramsar Wetland of International Importance in the USA and the first and only one in Ohio. Because of [the wetland's] small size (20 ha) [50 acres] the pipeline would have been an ecological and hydro-logical disaster in the long run.

Hooray!!!!!!!!! Thanks to all of you who helped in this incredible win for teaching, research, and nature!!!

Prof. Bill Mitsch, ORWRP Director

Posted on the Facebook page: "Save the OSU wetlands"

7.1 Why wetlands?

Erik Ness (2006), a freelance writer from Wisconsin, described our campus wetlands as "engines of ecological integrity" in an interview he was hired to do with me by the University of Notre Dame, where I received my bachelor's degree in 1969. Ness's article is reprinted in Appendix 7.A of this chapter. Wetlands are shallow to intermittently flooded ecosystems that are more com-monly known by such terms as swamps, bogs, marshes, and sedge meadows. While they are now revered and protected as important components of the natural landscape because of their capabilities in cleaning and retaining water naturally, preventing floods, and providing habitat and food sources for a wide variety of plant and animal species, it is estimated that more than half of the original wetlands in the lower 48 states have been lost to drainage and human development projects (Mitsch and Gosselink, 2015). In fact, Ohio has lost about 90% of its original wetlands, second in the USA only to the 91% loss rate of wetlands in California.

When we lose wetlands, we lose their ability to provide clean water, prevent floods, and enhance biological diversity. Many professional and civic orga-nizations call for the creation and restoration of wetlands to clean up our streams and rivers, and for river and floodplain restoration to recover lost habitat and other ecosystem services. Five million acres of wetlands in the Mississippi River Basin have been suggested as necessary to help prevent the "dead zone," or hypoxia, in the Gulf of Mexico. Large wetland restorations, at costs of billions of dollars, are underway throughout the world, for example, in the Florida Everglades, Louisiana Delta, and Mesopotamian Marshlands. Coastal wetlands are needed more than ever for protection from disasters such as the 2004 Indian Ocean tsunami and the 2005 Hurricane Katrina in New Orleans. Wetlands may indeed be the linchpin in fighting against climate change because of their major storage of carbon.

The U.S. Army Corps of Engineers oversees a regulatory program that results in tens of thousands of acres of wetlands being restored and created each year to replace wetlands lost to development. A report published by the National Academy of Sciences (Lewis et al., 1995) described in Chapter 5 concluded that much more research is needed before we can be assured that those wetlands that are constructed to replace wetlands destroyed for development can be successful. Even though a controversial Federal report suggested that there was even a net gain of wetlands in the United States from 1998 to 2004, the question of whether we can create and restore sustainable rivers and wetlands remained unanswered. I am optimistic that we can be successful, but only if there is a significant number of trained ecological engineers with appropriate backgrounds in hydrology, soil science, and botany along with engineering know-how.

I moved to The Ohio State University (OSU) in January 1986 as an Assistant Director of the School of Natural Resources. This position offered a significant increase in prestige, but there were two aspects for which my friends immediately questioned my sanity. First, I was entering, perhaps too early in my career, into a university administration position where I might be trapped forever and totally cut off from a research career; and second, Ohio had drained a higher percentage of its wetlands than any other state except California. It was not, therefore, a good place to develop a career in wetland science. I served as Assistant Director of the School of Natural Resources for one year, after which a new Director was chosen for the School and I could gently slide into a regular faculty position. Due to the lack of wetlands in Ohio, I decided to get into the business of creating and restoring new wetlands in Ohio and the Midwest. I had concluded that there could someday be a gigantic "market" for ecological engineering and science of wetlands in Ohio.

7.2 History of the Olentangy River Wetland Research Park
7.2.1 What came first

While I was still at the University of Louisville, I began collaborating with a leader in riverine ecological engineering, the recently deceased Dr. Donald L. Hey, then Director of Wetlands Research, Inc. in Chicago. Dr. Hey was constructing a new riparian wetland research park north of Chicago called the Des Plaines River Wetland Demonstration Project. It was an initiative that included riverine wetlands fed by pumps connected to the Des Plaines River. That collaboration strengthened when I moved to Ohio State (which was a little closer to Chicago than was Louisville) and I began to recruit good M.S. and Ph.D. students almost immediately (see Chapter 1). Thus, before the

Olentangy River Wetland Research Park (ORWRP) became a reality, I had already completed several OSU graduate students who did their research in Lake County, Illinois at or near the Des Plaines River Wetlands. That research at the Des Plaines River site was perfect training for refining our design for a floodplain wetland park in central Ohio. We learned, for example, that boardwalks were essential to minimize trampling the wetlands to death with trails everywhere. Some of the history of our lab's involvement in the Des Plaines River Wetland Demonstration Project is summarized in Chapter 6 of this book.

7.2.2 An explosion of media attention

At that same time, I had begun to investigate whether it would be possible to build a riverine wetland park at or near The Ohio State University, since my grad students doing research at the site north of Chicago had to drive hundreds of miles from Columbus to reach that project. Fortunately, my idea of a campus wetland was picked up by College of Agriculture media expert Judy Kauffeld (1991), who wrote a story in a relatively new magazine, *Ohio 21,* entitled "A New Swamp Dawning" with a cover title "A Swamp Is Born: Creating Wetlands Through Ecotechnology" (**Figure 7.1**).

A reporter spotted the *Ohio 21* magazine story and published a story on the front page of the *Columbus Dispatch,* entitled "Swamp may be next on OSU campus" (**Figure 7.2**) on May 9, 1991, right under a picture of General Norman Schwarzkopf, as the Gulf War in Kuwait was then winding down. The publication of that article came as a surprise to me, and it even included a simple sketch of four floodplain wetlands attached to the Olentangy River that I had sent them. The site proposed in my sketch was a former agricultural experimental field known as "Dodridge Bottoms," at the northern tip of the Columbus campus. So now we had a campus site to at least propose.

The media was not finished with this crazy idea of a professor wanting to build a swamp on a college campus. Less than two weeks later, a reporter from the *Wall Street Journal* ran a somewhat humorous article, "He's Battling for a Campus Where Students Can Soak Up Knowledge" (**Figure 7.3**), on the front page of the paper's Marketplace section. The university administrators who were still recovering from the news in the *Columbus Dispatch* story just two weeks earlier were now learning that this professor with a campus wetland goal was persistent. Ah, but where will the professor get the hundreds of thousands of dollars to make this wetland park happen? Certainly, the University would not fund it. In fact, a representative from the College of Agriculture came to one meeting early in this sequence and told me exactly that!

Luckily, an alumnus of The Ohio State University and president of a solid-waste company in central Ohio also read the story in the *Wall Street Journal* and immediately connected a big wetland regulation issue that he faced on

Figure 7.1 Cover of Ohio 21 *magazine with first press story about the Olentangy River Wetland Research Park idea written by Judy Kauffeld and entitled "A New Swamp Dawning" with a cover title "A Swamp Is Born: Creating Wetlands Through Ecotechnology."*

(Permission provided by Judy Kauffeld, College of Food, Agricultural, and Environmental Sciences, The Ohio State University.)

The Columbus Dispatch

SOME SU
High 74/Low 56
Details/12D

THURSDAY, MAY 9, 1991

35 Ce

Swamp may be next on OSU campus

Prof floats plan to build an Olentangy wetland

By Tim Doulin
Dispatch Staff Reporter

William J. Mitsch is on solid ground at The Ohio State University, but he longs for the earth to be wet and mushy.

A wetlands ecologist at OSU, Mitsch has proposed building a swamp on campus along the Olentangy River near Dodridge Street. It would be modeled after wetlands along the Des Plaines River north of Chicago, considered the premier experimental wetlands in the country.

"We are sending graduate students there to do research," said Mitsch, a professor in the School of Natural Resources.

"It occurred to me that we need a similar system here with our expertise to experimentally manipulate ecosystems."

Researchers could monitor water levels, soils, plants and the ecological process to find ways to increase the effectiveness of wetlands, Mitsch said.

Wetlands serve as ecological filters, removing pollutant-laden sediments from water that find their way into rivers and streams.

"Wetlands are sort of kidneys of the landscape," Mitsch said.

Ohio has lost about 95 percent of its original wetlands, often forced to make way for developments.

"We are trying to get the rivers and wetlands back together again. I think that would go as far toward cleaning up some of our streams as all the sewage treatment plants we build," Mitsch said.

The Olentangy River could stand to be cleaned up, but the proposed Buckeye Swamp — which would be up to 40 acres — would not be nearly big enough to do the job, Mitsch said.

"You would need thousands of acres to do that because it is a real polluted stream," Mitsch said.

"Our idea would be to demonstrate that you could make it work on smaller scales. But I guarantee that the water that comes out of the river into the wetlands will return to the river cleaner."

The proposed site is a flood plain, making it perfect to serve as wetlands. The river would be a source of water, and flooding would be welcome.

"They used to plant crops down there but said it is too wet to plant. Well, that is exactly what we want," Mitsch said.

In most cases, water depths would be controlled by researchers. A pump installed near the Olentangy would distribute water from the river to the wetlands, and a constructed swale would allow the water to eventually flow back to the river.

The wetlands would be stocked with plants and some vegetation would grow naturally. Wildlife would be attracted to the area.

The wetlands would cost about $200,000 to build, but it's possible money could come from a private source.

Wetlands are protected under the Clean Water Act and developers often are required to replace wetlands that they destroy. OSU could supply a happy solution, Mitsch said.

Mitsch's proposal must be reviewed by OSU's Office of Campus Planning and Space Utilization, which would forward a recommendation to the president's office.

No action has been taken on the proposal yet.

There are other potential sites on campus for wetlands, including land near the Fawcett Center for Tomorrow. A donor has offered land near the Mad River in western Ohio as a possible wetlands site.

"I think for the sake of doing research, we need to put something on campus," Mitsch said.

"I think this is different than putting one out in the middle of a cornfield."

Proposed wetlands

Floodplain boundary

Pump house

Wetland 1 | Wetland 2

Wetland 3 | Wetland 4 | Return flow to river

Separate cells for water level changes | To campus

Dodridge St.

An artificial wetland complex proposed for a site near OSU would allow biological research.

Source: William J. Mitsch, OSU — Dispatch graphic

Figure 7.2 Columbus Dispatch *story on May 9, 1991, as the first public announcement of a plan to build a wetland at The Ohio State University.*

his solid-waste disposal sites with what we were doing—trying to build successful wetlands that could mitigate wetland losses and do so at reasonable costs. Bottom line, I received a phone call from his attorney followed by the first of several checks as donations to a newly created university wetland fund account that still is listed today at Ohio State University. That first $300,000 total donation was sufficient to build two kidney-shaped wetlands, each 2.5 acres, and a river intake and pump system that could deliver known amounts of water depending on our experiment to each of the two wetlands. The system was similar to the preliminary sketch I provided to the *Columbus Dispatch* and shown in **Figure 7.2**, but with only two wetland basins rather than four.

7.2.3 The vision

The Olentangy River Wetland Research Park, 20 years after these press stories predicted that it might be next to impossible to create, became a real 50-acre campus facility located in central Ohio, USA, designed to provide teaching, research, and service related to wetland and river science and ecological engineering. At the "research park," scientists sought to understand: 1) how wetlands, rivers, and watersheds function and 2) if and how we can restore these

THE WALL STREET JOURNAL.

© 1991 Dow Jones & Company, Inc. All Rights Reserved

★★ Midwest Edition   TUESDAY, MAY 21, 1991   Bowling Green, Ohio

TUESDAY, MAY 21, 1991 **B1**

He's Battling for a Campus Where Students Can Soak Up Knowledge

By James S. Hirsch
Staff Reporter of The Wall Street Journal

Professors are sometimes criticized for studying soft sciences, but William J. Mitsch's science is downright mushy.

Mr. Mitsch, a professor of natural resources at Ohio State University, wants to build a 30-acre swamp just north of the Columbus campus. School officials bristle at the word "swamp"—they prefer "experimental wetlands"—but Prof. Mitsch has already dubbed the bog the Buckeye Swamp. It would be the site of important ecological research, but some students are already imagining the worst.

"I don't see it being in any of our pamphlets: 'Come see our swamp,'" says Jim Oliphant, a law student. "Wouldn't there be some kind of gnat problem, or mosquito problem? I think of a dark, dank morass where light doesn't go."

Actually, swamps aren't nearly that creepy. The wetlands wouldn't be a mosquito magnet, Mr. Mitsch says, and they would attract relatively benign animals—muskrats, beavers and waterfowl.

Prof. Mitsch envisions several interconnecting swamps on a flood plain near the Olentangy River, on university-owned property. He believes the project would help unlock the mysteries of the marshes: For instance, students could study how swamps break down impurities in water.

"That's the beauty of having it on campus," Prof. Mitsch says. "We can watch it for the next 100 years."

While many man-made swamps exist, building this one wouldn't be easy. It would require digging a shallow basin where the soil won't absorb the water. The basin would be seeded with plants—cattails, sedges and the like—then filled with water from the adjacent river.

Mr. Mitsch figures the whole shebang would cost about $200,000, but university officials say that's way too low. The school praises the idea but says it has no plans to fund the project. The site has other problems, such as a bike path that the city is planning to put in next to the project.

Nonetheless, Mr. Mitsch is determined not to let his swamp get bogged down. He says he'll try to raise the money from private industry and the government. His pitch will be that the country has lost millions of acres in wetlands, mostly because of land development, and scientists need to learn how to create new ones more efficiently.

Mr. Oliphant, the law student, says he sees a possible benefit: "If you had a decent swamp, I'd put a boat on it and bring a picnic lunch and a date."

Figure 7.3 Wall Street Journal *article on May 21, 1991, about a campus wetland proposed for The Ohio State University.*

systems. The 20-hectare (50-acre) site became a long-term, large-scale riverine wetland "research park" and teaching laboratory. There was no other facility of its kind on any other university campus in the world, so it also had as its mission the global dissemination of wetland science and ecological engineering.

The wetland research park has also served as a nature "park," providing habitat for a diversity of plants and animals that urban residents of Columbus and central Ohio might observe and enjoy. It was indeed possible to have a first-rate "living laboratory" that is also appreciated for its ecology and aesthetics in an urban region. Cooperation between the university and its urban neighbors became both symbolic and real at the Olentangy River Wetlands.

7.2.4 OSU's wetland park development 1990–2012

The research park was developed over 23 years in several phases:

- Phase 1 (1990–1994)—Construction of two experimental wetland basins and their water delivery system. Groundbreaking ceremonies and photo-ops for the two 1-hectare (2.5-acre) kidney-shaped deep-water marshes and a river water delivery system occurred in 1991–1992 (**Figures 7.4** and **7.5**). Pumps were installed on the northern tip of the floodplain site to bring water from the Olentangy River to the wetlands. Pumping officially began on March 4, 1994, the proclaimed birthday of these two kidney-shaped marshes (**Figure 7.6**).

Figure 7.4 *Early presentation of wetland park design to OSU President Gordon Gee April 1991. Left to right, Mohan Wali, Director, School of Natural Resources, OSU; Gordon Gee, President, The Ohio State University; Chris White, donor, Mid-American Waste, Inc.; Bill Mitsch, Olentangy River Wetland Park, Director, OSU.*

(Permission provided by University Archives, The Ohio State University.)

Figure 7.5 Groundbreaking team for the wetland basins, spring 1992. Left to right: Bill Mitsch, Director, Olentangy River Wetland Research Park, OSU; Chris White, donor, Mid-American Waste, Inc.; Gordon Gee, President, The Ohio State University, Columbus; Bobby Moser, Dean, College of Agriculture, The Ohio State University; Mohan Wali, Director, School of Natural Resources, The Ohio State University.

(Permission provided by University Archives, The Ohio State University.)

Figure 7.6 First addition of pumped water to one of the two 2.5-acre kidney-shaped wetlands, March 4, 1994.

(Photo and permission provided by William J. Mitsch.)

Figure 7.7 *Early planted vegetation developing in the "planted" kidney wetland basin.* **(Photo permission provided by William J. Mitsch.)**

Beginning in March 1994, river water continued to be pumped, day and night, into the two wetlands, except for drawdowns during pulsing experiments, for more than 20 years. The water then flows by gravity back to the Olentangy River through a swale and constructed stream system. In May 1994, one wetland basin was planted with marsh vegetation typical of wetlands in the Midwest; the other remained as an unplanted control. "Unplanted" did not mean "unvegetated." That basin was often more productive than the planted basin (Mitsch et al., 1998, 2012, 2014; **Figure 7.7**).

- Phase 2 (1994–1999)—Development of additional research, teaching, and outreach infrastructure at the site, including boardwalks (**Figure 7.8**), experimental mesocosms, a plant-material greenhouse, additional wetlands, instrumentation for long-term research, and a visitor's pavilion (**Figures 7.9** and **7.10**) made possible because of funds provided by John and Tana Sandefur of both Columbus, Ohio, and Sarasota, Florida. I had met the Sandefurs at an Ohio State University fundraiser in the mid-1990s in Naples, Florida, called "Winter College." I thought at the time that southwest Florida might be a nice place to retire to someday.
- Phase 3 (2000–2003)—Development and construction of the Heffner Wetland Research and Education Building on the site (**Figure 7.11**). Initial funding for the $3 million building began with $1.2 million

Figure 7.8 *Boardwalk construction occurred during the winters of 1995 and 1996 to minimize disturbance of vegetation and wildlife habitat.*

(Photo property of and permission provided by William J. Mitsch.)

from two Hayes Investment Fund grants that I received from the Ohio Board of Regents. These grants were the result of the efforts of a consortium of five partner Ohio institutions—Ohio State, Wright State, Shawnee State, Youngstown State, and Kenyon College—appealing for a wetland center that they could all use. Most of the remaining support for the building came through generous donations, particularly from Mr. Bill Heffner, for what became named the Heffner Wetland Research and Education Building at the wetland park. The decision to go forward with building construction was made on December 13, 2001, with construction beginning in spring 2002, and staff and students moving into the building from Kottman Hall on March 6, 2003.

- Phase 4 (2003–2012)—International collaborations, wetland endowments, and urban ecotourism. A $1.5 million endowment, named for a generous donor, Wilma H. Schiermeier, was established in 2004 to provide support in perpetuity for the Olentangy River Wetlands. A bike path shelter, named the AEP Bike Path Shelter (**Figure 7.12**), was constructed in 2006–07 adjacent to the city's bike path that passes through the ORWRP, and dedicated in September 2009, with the Mayor Mike Coleman of Columbus presiding over the dedication. He likened the facility—where we provided an air pump powered by solar collectors on the roof for bikers to pump air in their bicycle tires—to a "bike path service station."

Figure 7.9 *Ground floor of John and Tana Sandefur Wetland Pavilion that was fre-quently used to host evening "Moonlight on the Marsh" lectures for the general public.*

(Photo permission provided by William J. Mitsch.)

Figure 7.10 *A spotting scope was installed in the upper level of the Sandefur Pavilion in 1999 to enhance wildlife observations.*

(Photo permission provided by William J. Mitsch.)

Figure 7.11 *The Heffner Wetland Research and Education Building, named after Bill Heffner because of a generous donation from the Heffner estate, was constructed in 2002 with a move-in time of March 2003. The building featured modern labs, a mud room, a YSI data management center, a corner-windowed conference room with a singular view of the wetlands, and faculty, staff, and graduate student offices. This photo of the Heffner Wetland Building is from summer 2003.*

(Photo permission provided by William J. Mitsch.)

Figure 7.12 *The AEP bike path shelter was constructed in 2008–09 adjacent to the City of Columbus bike path and the bottomland hardwood forest along the Olentangy River. The bike path, originally scheduled to be built as a 40-ft wide easement through the bottomland forest, was rerouted after a suggestion by us to use non-forested land adjacent to the bottomland hardwood forest, saving thousands of trees.*

(Photo permission provided by William J. Mitsch.)

7.3 International recognition

My team and I continued to strengthen international networks of the ORWRP. That networking received significant recognition in 2008 when the Olentangy River Wetland Research Park was named the USA's 24th Ramsar Wetland of International Importance (**Figure 7.13**). To this day, it remains the only Ramsar Wetland of International Importance in Ohio and remains one of only 41 Ramsar sites in the USA. By comparison, the UK has 175 Ramsar sites, Canada has 37 Ramsar sites, and Mexico has 144 Ramsar sites.

In October 2010, the ORWRP received a Green Globe Award from Japan as the best wetland restoration site in North America. Additionally, Fulbright scholars from India and Estonia and visiting grad students from East China Normal University in Shanghai, China, and Ewha Woman's University in Seoul, Korea, completed major parts of their dissertation at the ORWRP.

CONVENTION ON WETLANDS
(Ramsar, Iran, 1971)

This is to certify that

Wilma H. Schiermeier Olentangy
River Wetland Research Park

has been designated as a

Wetland of International Importance

and has been included in the
List of Wetlands of International Importance
established by Article 2.1 of the Convention.
This is site No.: 1779

Secretary General
Convention on Wetlands

Date of designation *18 April 2008*

Figure 7.13 *Certificate from the Ramsar Convention on wetlands indicating that the Wilma H. Schiermeier Olentangy River Wetland Research Park is formally a Wetland of International Importance, effective April 18, 2008.*

(Photo of Ramsar certificate by William J. Mitsch; Ramsar certificate provided by the Olentangy River Wetland Research Park and Ramsar Convention, Geneva, Switzerland.)

7.4 Teaching

As an example of teaching activity at the ORWRP, 55 university courses, involving 1,000 students from seven OSU Colleges/Programs and several other colleges and universities, used the ORWRP in 2010. Ohio State course topics included: physical geography (College of Social and Behavioral Sciences); natural history and wetland ecology (College of Food, Agricultural, and Environmental Sciences); civil engineering and ecological engineering (College of Engineering); ornithology and conservation biology (College of Biology, Math and Physical Sciences); and Art and the Environment (Wexner Museum). Other colleges making use of the site included Columbus State Community College, the University of Pittsburgh, The Pennsylvania State University, and Northeastern Illinois University. A total of 70 students completed dissertations or theses in the 22-year period from 1992 through 2014, a rate of three theses or dissertations per year. By the end of 2014, there were 80 graduate student and post-doc names included on the "ORW Wetland

Hall of Fame" plaque hanging in the "Mitsch Family Kitchen" in the Heffner Wetland Building. While most of the students were from OSU departments and programs, some were visiting students from Europe and Asia.

7.5 Tours/public seminars

From 1994 through 2010, the ORWRP hosted 1,750 wetland tours and presentations for almost 34,000 participants. Among the many organizations taking tours were the Wexner Art and Environment Program, Women in Engineering for the College of Engineering, Ohio Water Environment Association, and the Wuhan (China) to Ohio Summer Program. Additionally, because of the bike paths and walking trails, an untold number of area residents or visitors have been able to enjoy the benefits of this urban wetland.

7.6 ORWRP's impact on wetland science, ecological engineering, and regional development

The scientific and ecological engineering contributions of the ORWRP to the protection and enhancement of the world's water resources have been significant. During the development and operation of the ORWRP over its first 20 years, the wetland park provided an economic advantage to the University of more than $11 million in extramural grants and contracts, donations, and short course fees. Over $1.8 million in endowments were raised in the first two decades of the ORWRP that have been used to support student scholarships and graduate student stipends and to maintain the physical structures at the site in perpetuity. Other endowments are currently in the works at the ORWRP to continue research that has revolutionized the fields of ecosystem restoration and ecological engineering, particularly as they relate to rivers and streams. Research results have been published in hundreds of scientific papers, abstracts, theses, and dissertations.

Ecological improvements of the riverscape in Columbus over our 20 years (**Figure 7.14**) have already occurred due to the presence of the ORWRP. From 1991 to 2013, research at the Schiermeier Olentangy River Wetland Research Park focused on reconnecting rivers to their floodplains through the application of ecological engineering principles to that restoration. Those principles include the following: 1. Design ecosystems that have homeostatic capability to smooth out and depress the effects of strongly variable inputs such as river flooding. 2. Design ecosystems for pulsing subsidies wherever appropriate and possible. 3. Allow ecosystems to self-design. The principles have also been applied toward many river restorations throughout the world, including the Mississippi River Delta in Louisiana, the Delaware Bay, the Mesopotamian Marshland in Iraq, and the 3.0-million-km^2 (1.16-million-mi^2) Mississippi-Ohio-Missouri (MOM) River watershed (Mitsch et al., 2001).

Figure 7.14 *College of Agriculture Dean Bob Moser giving a presentation in the Heffner Building lobby in May 2010 supporting the extension of the Ramsar Convention designation, earned in 2008 for the Wilma H. Schiermeier Olentangy River Wetland Research Park (ORWRP), to include a much larger part of the urban watershed downstream through Columbus, Ohio, USA. We proposed an extension of an existing Ramsar site to include a 14-km (8.7-mile) Olentangy/Scioto Ecosystem Corridor (OSEC)—a "blue/green ecological highway"—from the Olentangy River Wetland Research Park through the restored river as it flows through the campus downstream to a newer riverine, Grange Insurance Audubon Center, on the Scioto River south of the city Center (Mitsch, 2014).*

(Photo permission provided by William J. Mitsch.)

7.7 Nightmare ends the dream

The end of my affiliation with OSU and my beloved wetland park resulted from the arrival of a new Ohio State administrator or two with agendas to encourage retirement by high-salaried professors, partially due to the economic slump of 2008–10. About that same time, the university was asked by a regional natural gas company if the gas company could use the wetland site to replace an old pipeline with a new 90-cm (36-inch) natural gas (methane) pipeline that would extend directly north to south under the wetland park.

I was invited and politely went to a university meeting discussing this pipeline project with university administrators and the gas company. I made it known that my position was "Hell, no!" on routing this pipeline under these wetlands, which were directly connected to the groundwater below. I solicited and received many local, national, and international letters addressed to University President Gordon Gee (who has provided a Foreword to this book) and/or the Board of Trustees that all basically said the same thing: "Don't you dare build this pipeline under this priceless wetland research park" (see Appendix 7.B). A Facebook (FB) page was created by students in 2011 under the name "Save the OSU Wetlands." At the peak of this pipeline crisis, there were more than 500 students and concerned citizens affiliated with this FB page.

Finally, on December 2, 2011, I was able to announce on this same "Save the OSU Wetland" Facebook page (see Preface to this chapter) that Ohio State had decided to prevent the gas company on campus from constructing the pipeline under the Olentangy River Wetlands. Later, a state board decided that, rather, an alternative route should be used along the nearby Olentangy River Road. We had fought and won the good fight, but by then I had been courted by and was heading to Florida, where I subsequently spent a nice 10-year professorship in a wonderful climate at my Everglades Wetland Research Park in south Florida. I simply erased the words "Olentangy River" from my Ohio wetland's name and replaced it with "Everglades" for our new home in Naples, Florida.

While this pipeline under the wetlands was being proposed and some of my OSU administrators were failing to defend the wetland site, I was contacted by a search committee from Florida Gulf Coast University in the spring/summer 2011 with an offer of an endowed chair and eminent scholar position. Given the difficult political situation that this pipeline proposal put me in at OSU and knowing the potential lasting damage the pipeline would have on the OSU wetlands if the University allowed them to build it, I said I was interested in the Florida offer.

In October 2012, I accepted FGCU's offer of the Sproul Endowed Chair and a Directorship of FGCU's new "Everglades Wetland Research Park." I took several of my OSU grad students with me to Florida. Indeed, they too had all been disheartened by the brazen pipeline threat allowed by some of OSU's leadership and were excited by the possibility of doing research in Florida's unique ecosystem, the Everglades.

References

Kauffeld, J. 1991. A new swamp dawning. *Ohio 21*, Vol 6, No. 1. (March 1991). College of Agriculture, The Ohio State University, Columbus, OH. 6–9 pp.

Lewis, W.M., B. Bedford, F. Bosselman, M. Brinson, P. Garrett, C. Hunt, C. Johnston, D. Kane, A.N. Macrander, J. McCulley, W.J. Mitsch, W. Patrick, R. Post, D. Siegel, R. W. Skaggs, M. Strand, and J.B. Zedler. 1995. Wetlands: Characteristics and Boundaries. The National Academies of Science Press, Washington, DC.

Mitsch, W.J. 2014. Unifying a city with its natural riverine environment for the benefit of both: Extending Ohio's only Ramsar Wetland to a much larger river ecosystem corridor. *Ecological Engineering* 72: 138–142.

Mitsch, W.J., J.K. Cronk, and L. Zhang. 2014. Creating a living laboratory on a college campus for Wetland Research—The Olentangy River Wetland Research Park, 1991 to 2012. *Ecological Engineering* 72: 1–10.

Mitsch, W.J., J.W. Day, Jr., J.W. Gilliam, P.M. Groffman, D.L. Hey, G.W. Randall, and N. Wang. 2001. Reducing nitrogen loading to the Gulf of Mexico from the Mississippi River Basin: Strategies to counter a persistent ecological problem. *BioScience* 51: 373–388.

Mitsch, W.J. and J.G. Gosselink. 2015. *Wetlands, 5th ed.* John Wiley & Sons, Inc., Hoboken, NJ. 744 pp.

Mitsch, W.J., X. Wu, R.W. Nairn, P.E. Weihe, N. Wang, R. Deal, and C.E. Boucher. 1998. Creating and restoring wetlands: A whole-ecosystem experiment in self-design. *BioScience* 48: 1019–1030.

Mitsch, W.J, L. Zhang, K.C. Stefanik, A.M. Nahlik, C.J. Anderson, B. Bernal, M. Hernandez, and K. Song. 2012. Creating wetlands: Primary succession, water quality changes, and self-design over 15 years. *BioScience* 62: 237–250.

Ness, Erik. 2006. Waist-deep in ecological integrity. *Notre Dame Magazine*, Spring 2006. Notre Dame, IN.

Appendix 7.A

Text of an editorial written by Erik Ness in 2006 entitled *"Waist-deep in ecological integrity"* for my undergraduate alma mater Notre Dame in *Notre Dame Magazine*, Spring 2006. Notre Dame, Indiana (Permission to Reprint provided by the author Eric Ness and Notre Dame Magazine, Notre Dame, Indiana). (Please refer to page 92.)

Appendix 7.B

Selected letters of support for *ORWRP against the Columbia Gas pipeline*, October 2011; Similar letters were received, all opposing construction of a pipeline underneath the Olentangy River Wetland Research Park, from 17 other sources (permission has been provided by the current officers of U.S. National Ramsar Committee; Dr. Jerry Pausch, Chair, Olentangy River Wetland Research Park Advisory Committee; and Mr. Mike Peppe, Naples Florida and Columbus Ohio, to publish these three letters). (Please refer to page 96.)

Waist-deep in ecological integrity

By Erik Ness

Notre Dame Magazine Spring 2006

Bill Mitsch '69, a professor of natural resources and environmental science at Ohio State University, hit a political nerve when he spoke up about how heavy human hands on the Mississippi River had exacerbated the problems in New Orleans.

"It makes no sense to spend any money rebuilding the city without addressing the long-overdue issue of restoring the natural features that once protected New Orleans from storms," Mitsch told the *Columbus Dispatch*. He called the extent of the damage to the river and related wetlands "criminal."

Swamps, bogs, marshes, fens, quagmires—whatever you call them, Mitsch says wetlands are the salvation of the world. We need more of these engines of ecological integrity, and he knows how to build them. "They've really got to let that river go, left and right," he says. "We get in trouble when we're arrogant enough to think that we are in total control."

As a young man with a degree in mechanical engineering, Mitsch was fascinated with power plants and thermodynamics. He wound up at Commonwealth Edison, which lights up Chicagoland. He eagerly watched the environmental movement take root in the political ferment of the day then realized uncomfortably that he was working for a major polluter. After a stint in the company's new office of environmental affairs, he left for graduate school at the University of Florida.

Among his mentors at Florida was H.T. Odum, one of the premier ecologists of the last century. A big thinker and master showman, Odum would astonish companions when, donned in street clothes, he would stride into the water without interrupting his conversation.

Odum needed a graduate student for a unique experiment: What happens when you pumped processed sewage into a parched cypress swamp? Could the water revive the swamp, and would the swamp then finish cleaning the water? While scouting the experimental site, Odum tapped Mitsch for his fellowship of mud. "It was mucky and muddy and messy," recalls Mitsch. "[Odum's] just having a ball, and he turns to me and says 'You want to do this stuff?' And I said 'Hell yes.'"

In the South, Mitsch says, "You see how powerful ecology is." When the experiment worked, the possibilities set his mind ablaze. "It all clicked: all the cylinders of my life, from understanding thermodynamics to learning about ecology," he says.

Since Europeans had arrived in North America, wetlands had been seen as wastelands: something to drain, fill, cultivate or excavate. From 1780 to 1980 the lower 48 lost on average an acre of wetland every minute. But wetlands also were emerging as a flash-point where ecology and economics collided. Scientists argued that wetlands provided real economic benefits: They filter pollutants from water, buffer us from floods, and nurture a bounty of fish and wildlife.

Wetland policy became one of the most contested arenas of environmental policy. Businesses and communities often found themselves regulatory hostages to an inconveniently located stretch of damp ground. One possible solution was creating or restoring wetlands to compensate for those destroyed through development. But debate raged: Are wetlands created by people as good as natural wetlands?

Answering this question has been Mitsch's passion. At Ohio State, under-used fields near the Olentangy River were secured. Restoring old wetlands is usually easier than cutting them from whole cloth, but these had never been wetlands. Mitsch appreciates the exquisite and complex beauty of natural wetlands tended by Mother Nature over eons but argues that wetland function is mostly a matter of plumbing. "If you get the hydrology right, you can have wetlands on the moon."

He and his students designed two kidney-shaped catchments, and in 1994 they were filled with water siphoned from the Olentangy. One basin was planted with a half-dozen local wetland species; the other was allowed to follow nature's course. Within three years the two had converged and were nearly identical in terms of plant species and water filtration. Such ecological benefits as water purification and nutrient sequestration are similar in both wetlands.

Because we can create a wetland, says Mitsch, "We need to do stuff on a much bigger scale if we really expect wetlands to make a difference." Among his more radical proposals is to tackle the dead zone in the Gulf of Mexico with a minimum of 24 million acres of new and restored wetlands in the Mississippi River basin.

He has some practice with wetland work on a big scale. In Iraq, Saddam Hussein attempted to strangle resistance by draining massive wetlands in the south; Mitsch has twice traveled to the region to teach restoration techniques.

Using natural systems on this grand stage is an emerging field called ecological engineering, and Mitsch, with Danish colleague Sven Erik Jorgensen, wrote a pioneering text. In 2004 their collective works netted the authors the coveted Stockholm Water Prize, which comes with a $150,000 cash award. "The 21st century will be the century of repairing the planet," says Mitsch. "There needs to be a profession that knows how to do it right."

***

Erik Ness writes about science and the environment from his home in Madison, Wisconsin.

US National Ramsar Committee
http://www.ramsarcommittee.us/

c/o Professor Kim Diana Connolly
2011-2012 Chair
University at Buffalo Law School
519 O'Brian Hall, North Campus
Buffalo, NY 14260-1100
716/645-2092
kimconno@buffalo.edu

3 October 2011

Jay D. Kasey
Senior Vice President for Administration and Planning
The Ohio State University
101 Bricker Hall
190 N. Oval Mall
Columbus, Ohio 43210 USA

Re: Raising Concerns re Proposed Gas Lines under the Internationally-Designated Olentangy River Wetland Research Park

Dear Vice President Kasey,

I am writing to you today in my capacity as the Chair of the United States National Ramsar Committee (USNRC), an organization dedicated to supporting the goals and objectives of the Ramsar Convention on Wetlands within the United States and internationally. The USNRC and Ramsar promote the conservation and wise, sustainable use of domestic and international wetlands.

The Ramsar Convention on Wetlands is an intergovernmental treaty adopted on February 2, 1971. Over 150 countries, including the United States, are parties to the Ramsar Convention. One of the primary obligations of a Ramsar party is to designate sites as "Wetlands of International Importance." Amazing wetlands worldwide have been designated as wetlands of international importance; only 30 sites in the United States have earned this prestigious designation. As you know, one of those sites is the Wilma H. Schiermeier Olentangy River Wetland Research Park, the 24th Ramsar Wetland of International Importance in the USA and the only one in the state of Ohio. This designation was earned only after an extensive scientific review of the site, and a rigorous application process both within the United States and at the Ramsar

Secretariat in Switzerland. This is one of the first created wetland sites on the globe to receive a Ramsar designation, based on the diverse flora and fauna at the site, the extensive on-site wetland research, and the significant ecotourism at this urban wetland (approximately 150 tours per year are given to thousands of visitors).

The USNRC was notified recently of your University's consideration of a request to permit construction of a high-pressure gas pipeline under the entire length of the Wilma H. Schiermeier Olentangy River Wetland Research Park. The OSU Ramsar wetland has internationally important ecology, a global research and teaching reputation, and vibrant outreach to the community. I thus wanted to write to you and clarify obligations pursuant to the Convention.

The Ramsar Convention's mission is "the conservation and wise use of all wetlands through local and national actions and international cooperation, as a contribution towards achieving sustainable development throughout the world." The wise use of wetlands is defined as "the maintenance of their ecological character, achieved through the implementation of ecosystem approaches, within the context of sustainable development."

Once a site is designated, under Article 2 of the Convention, Contracting Parties agree that they "shall formulate and implement their planning so as to promote the conservation of the wetlands included in the List," and further direct that "each Contracting Party shall arrange to be informed at the earliest possible time if the ecological character of any wetland in its territory and included in the List has changed, is changing or is likely to change as the result of technological developments, pollution or other human interference." The full text of the Convention is available at http://www.ramsar.org/cda/en/ramsar-documents-texts-convention-on/main/ramsar/1-31-38%5E20671_4000_0___ .

At the most recent international Conference of the Parties in 2008 in Korea, attended by Bill Mitsch of OSU (as an internationally recognized wetland expert), the delegates adopted a Declaration that recognizes "the urgent need for governments, international organizations, the private sector and civil society to understand more fully the roles they can and should play in securing the future health of wetlands and the maintenance of their ecological character, in relation to the global commitments made under the Ramsar Convention..." A copy of that Declaration is attached.

The USNRC understands the difficult decision that OSU is facing with regards to permitting this underground easement on your campus. This letter is intended to remind you of the recognized ecological treasure on your campus, and suggest you consider the construction of this pipeline in the context of the Ramsar Convention obligations.

From my perspective beyond USNRC, I too work and serve as an administrator within a state university system. I understand the pressures of this current economic time, as well as the realities of the many stakeholders that need to be satisfied within a university system. Yet on behalf of the USNRC I urge you recognize the ramifications of taking an action with possible short-term advantages but almost certain long-term detriment to such an internationally significant treasure.

I would be happy to discuss the Ramsar Convention and its obligations in more detail with you. Feel free to contact me at 716/645-2092 or kimconno@buffalo.edu.

Very truly yours,

Professor Kim Diana Connolly
Chair, USNRC

Attachment: *Changwon Declaration*

cc: Terry Foegler Associate Vice President Physical Planning, The Ohio State
            University
        Mike Mitchell, Assistant Vice President/Associate Gen. Counsel, The Ohio State
            University
        E. Gordon Gee, President, The Ohio State University
        William Mitsch, Director, Olentangy River Wetland Research Park, The Ohio State
            University
        Krishna K. Roy, Branch Chief, Global Programs, Division of International
            Conservation, U.S. Fish and Wildlife Service, Department of the Interior
        Nick Davidson, Deputy Secretary General, Ramsar Secretariat

Olentangy River Wetland Advisory Committee

Dr. Jerry B. Pausch, Chair
P.O. Box 413
Leesburg, Ohio 45135
Email: jbp3131@yahoo.com

October 6, 2011

Mr. Jay D. Kasey
Senior Vice President for Administration and Planning
The Ohio State University, 101 Bricker Hall
190 N. Oval Mall, Columbus, Ohio 43210

Dear Mr. Kasey,

As long-time donors to The Ohio State University, and especially to The Wilma H. Schiermeier Olentangy River Wetland Research Park on The Ohio State University campus, we, the signatories to this letter, are very concerned by reports that the University and Ni-Source/Columbia Gas are in the process of permitting a high-pressure gas pipeline to be routed under the Wilma H. Schiermeier Olentangy River Wetland Research Park. This wetland research park is the 24th Ramsar Wetland of International Importance in the USA, an ecological and research treasure that is internationally significant and is the first and only Ramsar Site in the State of Ohio. The Wilma H. Schiermeier Olentangy River Wetland Research Park was designated as an International Ramsar Wetland in 2008 because of its unique ecology, consummate research achievements, teaching reputation, and outreach to the community, the nation, and the international community.

The Wilma H. Schiermeier Olentangy River Wetland Research Park has been designated by the U.S. Corps of Engineers as jurisdictional protected Waters of the United States under the Federal Clean Water Act, Section 404. We are concerned that the NiSource/Columbia Gas Corporation has chosen this location as their preferred path to install the pipeline, when alternative paths are available. The undersigned oppose the pipeline route in principle, and also because there has not been any guarantee that there will be no impact. The intended plan of action may significantly affect this world-renowned Ramsar Wetland of International Importance, a 50-acre, urban riverine woodland and wetland oasis of natural greenspace habitat in which over 174 bird species have been identified, and a site that serves as a classroom and laboratory for undergraduate and graduate students of The Ohio State University as well as a for a consortium of Ohio institutions of higher learning, and for area grade schools and high schools as well as birders and other nature enthusiasts.

To allow this high-pressure gas pipeline to be routed under the Wilma H. Schiermeier Olentangy River Wetland Research Park, which has been in operation for more than twenty years and has attracted donations in excess of 7 million dollars through 5,300 donations in support of innovative research, would be an international public relations nightmare for The Ohio State University. The Schiermeier Olentangy River Wetland Research Park is the prototype for research parks at other universities, not only in the United States but in many other countries as well. It is, in fact, the gold standard against which other university research parks are measured. The OSU wetland site attracts thousands of visitors, researchers, and students annually and in all seasons.

Ironically, the project is planned to commence in the same time frame as the Eco-Summit 2012, hosted by the MORPC, OSU, INTECOL, and SER, an event held previously in Copenhagen, Denmark; Halifax, Canada; and Beijing, China. A featured destination for Eco-Summit attendees will be The Wilma H. Schiermeier Olentangy River Wetland Research Park. To negatively impact the site of this premier research facility and damage or degrade the integrity of its hydrology and geology would contravene all research methods and results and seriously denigrate the sterling environmental stewardship reputation of The Ohio State University.

We are aware that the consultants of NiSource/Columbia Gas claim there is low risk from the installation of the high-pressure gas pipeline. There has been no "impact study" carried out by experts; thus, no risk is the only option. Baseline data of long-term research projects are threatened by intrusion of unnatural variables if this planned route proceeds. NiSource/Columbia Gas must pursue alternate routes.

Please respond to our appeal to Dr. Jerry Pausch at the address or email above.

Sincerely,

Jerry B. Pausch, Ph.D.
Chair, ORW Advisory Committee
Leesburg, OH
OSU President's Club

William H. Resch
New Albany, OH
OSU-M.A. Class of 1972

Ruthmarie H Mitsch, Ph.D.
Editor, *Research in African Literatures* and *Spectrum*
The Ohio State University

Bernard F. Master, D.O.
Worthington, OH
OSU President's Club

Karen S. Kelly
Upper Arlington, OH
OSU-B.S. Class of 1977

Edward Bischoff, P.E.
President, Bischoff, Miller & Assoc.
Powell, OH

Jared R. Nodelman
New Albany, OH
OSU President's Club

Thomas X. Singer
Wincheser, OR
OSU President's Club

Paul Sipp
Upper Arlington, OH

Walter W. Schiermeier
Cincinnati, OH
OSU-B.S. Class of 1962

cc:
Gordon Gee, President, The Ohio State University
Terry Foegler Associate Vice President Physical Planning, The Ohio State University
Mike Mitchell, Assistant Vice President/Associate Gen. Counsel, The Ohio State University
William Mitsch, Director, Olentangy River Wetland Research Park, The Ohio State University

MICHAEL G. PEPPE
2845 Margate Road
Columbus, Ohio 43221
614-457-6650, Office
614-314-1474, Cell

October 22, 2011

Professor William Mitsch, Director
OSU Wilma H. Schiermeier Research Park
Heffner Wetland Research Bldg.
352 W. Dodridge Street
Columbus, OH 43202

Re: *The Columbus Dispatch* article, "Pipeline Could Pass Under OSU Wetlands"

Dear Professor Mitsch:

I am frankly exasperated by the recently reported proposed alignment of a Columbia Gas high pressure gas line under The Ohio State University Wetland Research Park. I have been around The Ohio State University for a very long time and cannot think of one other <u>internationally</u> acclaimed university facility. This is the type of positive recognition that officials say they want for OSU especially when there is so much negative publicity nationally concerning the football program.

Public utilities share a community responsibility to make decisions based on fact finding. If such due diligence had been done during the evaluation portion of the planning process, this letter would not have been required.

The facts are glaringly apparent that The Ohio State University Wetlands Research Park has earned numerous national and international awards, culminating in the most prestigious international award that is an honor to OSU, Columbus and Ohio. And now there is a proposal to jeopardize this treasure by running a high pressure gas line under it? What are university and utility officials thinking?

This facility is one of a kind for an academic campus and brings Ohio State positive international recognition. It should be a point of pride, not some cheap commodity that can be so easily endangered by such a poorly conceived plan.

The Columbia Gas system needs to be commended for its overall safety record and vilified when decisions such as this pipe line alignment endanger an outstanding academic achievement within the national and international scientific community.

Columbia Gas and OSU should be embarrassed and ashamed to even release their spin doctors' comments that, "We want to do this right and make sure there is <u>no meaningful risk</u> to the wetlands." Oh, really!

Do the right thing and remove from consideration a Columbia Gas high pressure gas line underneath OSU's research wetlands of international significance.

Sincerely,

Mike Peppe

Michael G. Peppe

Cc: OSU Board of Trustees   Archie Griffin, OSU Alumni Assn.   Ohio Power Siting Commission
    G. Gordon Gee, OSU President   Ohio Governor John Kasich   Jack Partridge, Columbia Gas

Restoring the Florida Everglades

Professor Bill Mitsch in front of his Everglades Wetland Research Park, located at the Naples Botanical Garden in Naples, Florida.

DOI: 10.1201/9781003374619-8

8.1 Introduction

Florida's ecosystems, especially the wetland ecosystems of subtropical Florida, have fascinated me and held my interest since I enrolled in the University of Florida's Environmental Engineering Science Department (EES) in 1971. During my first academic year at U of F, I saw an announcement in Black Hall, the home of the EES Department, that graduate students could apply for a 12-credit hour course entitled "Man and Environment" to be held at the University of Leiden, The Netherlands, in the summer of 1972. It was a course co-organized by University of Florida and two Dutch universities. Graduate students who submitted successful proposals would have travel support to and from The Netherlands as well as lodging expenses covered in Leiden during the 8-week course. Little did I realize that this international course, which included the energy-based theories of Professor Howard T. Odum as part of the course content, was to be the first step in my transformation from being a power plant engineer to a systems ecologist and H.T. Odum student.

My dissertation, over 400 pages long, written under Odum, and completed in June 1975, was titled "Systems analysis of nutrient disposal in cypress wetlands and lake ecosystems in Florida" and involved ecological modeling of lakes and wetlands and work on a novel ecological engineering project that made use of cypress swamps and water hyacinth marshes as wastewater cleaning systems (Mitsch, 1975). It was one of the first such research projects on this subject in the country.

As part of my dissertation and our study utilizing cypress domes to improve water quality, H.T. Odum proposed that we use Corkscrew Swamp, a magnificent cypress slough in Naples, Florida, as a control site for our cypress domes near Gainesville and the University of Florida. It was a long drive from the university to Corkscrew Swamp in SW Florida, but when I first arrived there, I was blown away by the beauty of the swamp. Later on, those who managed Corkscrew referred to it as a sanctuary. It reminded me of an ecological cathedral, from the beginning, with stately cypress trees more than 500 years old and a beautifully diverse subcanopy of wetland plants and smaller wetland trees. Before I left Corkscrew Swamp on my first trip there, I firmed up my career objectives that I was going to be first and foremost a wetland scientist as well as an ecological engineer. I was in a wetland church!

I visited Corkscrew several times in my last two years in grad school for some sampling, but truthfully, just to be there. Little did I know that I would write five editions of my "wetland bible" (Mitsch and Gosselink, 2015) and further, Corkscrew Swamp personnel often agreed to post my book in their gift shop for sale. I also was thrilled when I got invited in the 2010s to give evening seminars there on several occasions. I guess that analogy made me a preacher in the cathedral.

My first 36 years after my Ph.D. at University of Florida as a professor were "up north" in Chicago, Louisville, and Columbus, a long-enough career for any professor. But when an opportunity was presented to me to return to the Florida ecosystems that I so appreciated 36 years earlier, I jumped at the chance. In fact, there was a double advantage, as the position was also in south Florida, where I had headed up some wetland research grants and served on wetland-related committees related to the Florida Everglades with the South Florida Water Management District (SFWMD). Welcome to the next-to-last chapter of my memoirs.

8.2 Water quality and the Florida Everglades

The restoration of the Florida Everglades, one of the largest wetland areas in the world, actually involves many separate initiatives being carried out in the 4.6-million-ha (11-million-acre) Kissimmee-Okeechobee-Everglades (KOE) region in the southern third of Florida (**Figure 8.1**; Mitsch, 2016). The basic plan aims at restoration that brings the KOE to something closer to the original hydrology of the region (**Figure 8.1a**) by sending less water from the upper watershed to the oceans via the Caloosahatchee River to the west and the St. Lucie Canal to the east and, thus, directing more of the water to the Everglades south of Lake Okeechobee (**Figure 8.1b**). The Kissimmee River, including the Upper Chain of Lakes, flows 166 km (103 miles) to Lake Okeechobee and is now subject to a major restoration to recreate a broad floodplain that had been dredged and straightened by state and Federal agencies in the 1960s. Lake Okeechobee, a 1890-square kilometer (730-square mile) body of shallow water, is the heart of this KOE system.

8.3 Working with water management contracts in Florida

Sometimes, the things I struggled for in my career had to do with protecting academic freedom, a sacred tenet in universities that now appears to be under fire everywhere. My return to Florida wetland research began simply enough. I already had a grant from Florida while I was at The Ohio State University, entitled "Assessing Nutrient Removal Efficacy and Uptake of Several Native Wetland Plant Communities," which was in effect from 2010 to 2013 and involved large-scale mesocosms (1 m × 6 m flow-through troughs) adjacent to some of the treatment wetlands (called STAs, or stormwater treatment areas, by the state of Florida) in the Florida Everglades.

Rick Scott was the newly elected Florida governor in 2010, just when our grant was starting. Scott, who appeared to have little interest in supporting environmental science or research, decimated the water management districts in Florida, cutting $700 million from their budgets in 2010–11. Those cuts have been suggested as one of the reasons why Florida experienced water

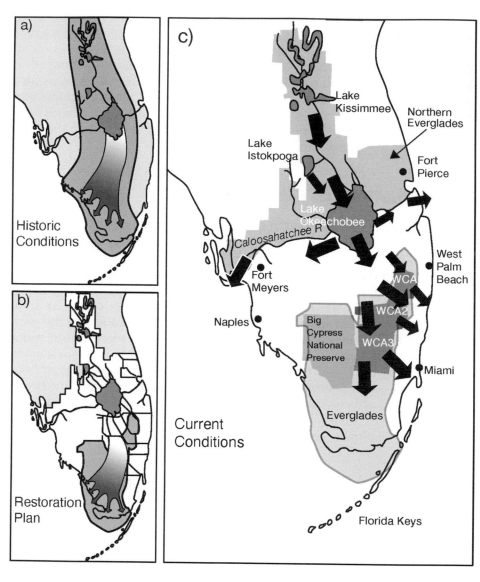

Figure 8.1 *Illustrations of Florida Everglades' current disturbed condition and restoration plan (from Mitsch and Gosselink, 2015; Mitsch, 2016): (a) historic conditions of the Florida Everglades hydrology in pre-settlement times; (b) the flow conditions desired when the Everglades is restored; and (c) typical flow conditions in recent decades in the Everglades where much of the water is sent east and west to the sea rather than south to the Everglades.*

(Permission for this figure from Mitsch and Gosselink, 2015 provided by John Wiley & Sons, Inc.)

quality crises beginning in 2018 (Geng, 2018; News Service of Florida, 2018). The damage that my research in Florida suffered because of Scott's cuts was serious. Even though my grant itself was not cut, my post-doc on our grant, Dr. Keunyea Song, was very nervous that we would in fact be shut down, and I could not disagree with her. So, I encouraged her to look for a backup position. As a result, she sought out and accepted another post-doc position, not in Florida but in the Midwest, and my research team was thereby virtually destroyed just as we were reaching the final year of report writing for our three-year project.

I gave an oral presentation about our research in April 2013 as required by our contract at the end of my first academic year at Florida Gulf Coast University (FGCU). Final report drafts were submitted in August 2013, with the final draft then submitted on December 31, 2013. The review process for this grant was like nothing I had ever experienced in my academic career. I was truly pulled into a state agency in turmoil. Good people were forced out of the agency because of these budget cuts.

As with all my funded research, I submitted the results to peer-reviewed journals, in this case *Ecological Engineering*, and with a good set of at least ten further professional reviews, it was published as Mitsch et al. (2015). The very agency that had funded us then hired academic and outside professionals to write rebuttals to our paper in the same journal.

I recall that when I received this Florida contract while I was still at The Ohio State University, the OSU research office expressed surprise that the state of Florida had a stipulation in the contract that they, as a state agency, had the right to change the grant reports if they did not agree with the findings of the principal investigator. This created a gigantic conflict of interest when they revise or rewrite reports to their benefit, trampling on the writers' academic freedom. In the end, our conclusions about which plant communities worked best in improving water quality were valid and reasonable conclusions.

8.4 Working for the friends of the Everglades on an EAA Reservoir review

In contrast to my experience described above, I was hired a few years later by a well-known non-governmental organization (NGO) in south Florida named the Friends of the Everglades. The Friends of the Everglades was started by Marjorie Stoneman Douglas, the author of the classic *The Everglades: River of Grass* (Douglas, 1947) and a long-time Florida Everglades protector. In 2017, The Friends asked me to review a new $2 billion Everglades water management project called "The EAA Reservoir," a plan designed to clean up Lake Okeechobee's polluted water and send it south to the Florida Everglades rather than continuing to send polluted water to Gulf of Mexico and Atlantic Ocean coastal waters (**Figure 8.1c**). This was a $2 billion project, so every consultant and environmental organization in south Florida was interested in it. I reviewed the project for the Friends of the Everglades and concluded that this proposal did not appear to me to have enough water quality treatment to guarantee that the water going to the Everglades would meet water quality standards acceptable for the Everglades.

Our conclusions are enumerated below:

1. While the proposed flow south to the Everglades from Lake Okeechobee **would increase by 76%** from 0.26 to 0.46 billion m³/year (69–121 billion gallons/year), only 2,630 ha (6,500 acres) of new treatment wetlands **(a mere 13% increase)** were proposed to clean the water. The system was going to be overloaded from the start. We estimated that 20,000–40,000 hectares **(50,000–100,000 acres) of treatment wetlands would be needed** to deliver clean water to the Everglades.

2. A 7-m (23-foot) deep 4,100-ha (10,100-acre) reservoir was to be built for seasonal water storage, but it was not clear that this reservoir would do much at all to improve water quality. It was mostly designed for water storage.

3. Further, while the estimated average concentration of total phosphorus flowing out of Lake Okeechobee at that time was 147 parts per billion (ppb) (Gary Goforth, personal communication), the average inflow to the current and future treatment wetlands south of the EAA is about 100 ppb. It is therefore probable that the phosphorus concentrations reaching current and future treatment wetlands south of the EAA would be too high and would threaten existing state and federal standards on Everglades water quality.

4. The new EAA Reservoir will not have any natural feature of aquatic ecosystems typical of those found in the Florida Everglades in ecology, morphology, or hydroperiod. It will therefore in no way represent a "restoration" of the Florida Everglades. The hydroperiods will be wrong and exaggerated for south Florida ecology—the way wetland hydroperiods shifted in the Great Lakes due to the creation of diked marsh hunt clubs and conservation areas (see Mitsch and Gosselink, 2015). The potential amplitude of the annual hydroperiods of up to 7.0 m (23 feet) in the EAA Reservoir is exceeded only rarely in natural or human-created rivers or wetland ecosystems, e.g., the Amazon River (Junk et al., 1989 or The Three Gorges Dam reservoir (Mitsch et al., 2008), let alone in the shallow-water Everglades. Thus, this EAA Reservoir may become a "freak ecosystem" over time, i.e., an aquatic ecosystem dissimilar in hydrology and probably ecology to any other aquatic ecosystem in the Everglades.

5. Most nutrient-rich lakes in our experience become occasional or even permanent sources, rather than sinks, of nutrients—Buckeye Lake, Ohio (W.J. Mitsch, personal experience), Taihu Lake in China

(Kelderman et al., 2005), and even Lake Okeechobee itself (Havens and James, 2005). It is highly probable that the EAA Reservoir will not be a nutrient sink in most years, although that is an assumption included in this plan.

8.5 History of this project after six years

After we submitted our report to the Friends, we published, with their permission, our results in a peer-reviewed paper (see Mitsch, 2019). Yet most organizations still seemed to believe that this $2 billion project would work as designed. It seemed that receiving $2 billion in funding from state and federal governments was more important than guaranteeing that the proposed EAA Reservoir would work. I frequently heard: "Well, the project is not perfect, but let's do it while the money is there." It is now 2023 and the EAA Reservoir has not been constructed; we started our investigation with the Friends of the Everglades six years ago.

Sometimes, time works in Nature's favor by filtering out bad ideas. That is hopefully what is happening here. There is a benefit in having the federal and state government jointly supporting such a gigantic project and thereby providing an additional level of safety for publicly financed water projects. In this case, the U.S. Army Corps of Engineers appears to be less excited than the State of Florida in supporting this project.

In summary, it was worth it to stay the course on my minority opinion that this EAA Reservoir project still needs to go back to the drawing board to make sure that only clean water is discharged into the Everglades. Money by itself is not the solution, but designing and constructing ecological engineered systems that will clean water before it enters the Everglades is essential. We may have saved the taxpayers in Florida and the USA a significant amount of money.

References

Douglas, M.S., 1947. The Everglades: *River of Grass*. Rinehart & Co., Inc., New York, NY.

Geng, L. 2018. Did Rick Scott cut $700 million from water management? *Politifact*. Retrieved August 27, 2018.

Havens, K.E. and R. T. James, 2005. The phosphorus mass balance of Lake Okeechobee, Florida: Implications for eutrophication management. *Lake and Reservoir Management* 21: 139–148.

Junk, W. J., P. B. Bayley, and R.E. Sparks. 1989. The flood pulse concept in river-floodplain systems. In D. P. Dodge (ed.) *Proceedings of the International Large River Symposium. Special issue of the Journal of Canadian Fisheries and Aquatic Sciences* 106: 11–127.

Kelderman, P., Z. Wei, and M. Maessen. 2005. Water and mass budgets for estimating phosphorus sediment-water exchange in Lake Taihu (China P. R.). *Hydrobiologia* 544: 167–175.

Mitsch, W.J. 1975. Systems analysis of nutrient disposal in cypress wetlands and lake eco-systems in Florida. Ph.D. dissertation, University of Florida, Gainesville, FL. 421 pp.

Mitsch, W.J. 2016. Restoring the Greater Florida Everglades, once and for all. *Ecological Engineering* 93: A1–A3.

Mitsch, W.J. 2019. Restoring the Florida Everglades: Comments on the current reservoir plan for solving harmful algal blooms and protecting the Florida Everglades. *Ecological Engineering* 138: 155–159.

Mitsch, W.J. and J.G. Gosselink. 2015. *Wetlands*, 5th ed. John Wiley & Sons, Inc., Hoboken, NJ. 744 pp.

Mitsch, W.J., J. Lu, X. Yuan, W. He, and L. Zhang. 2008. Optimizing ecosystem services in China. *Science* 322: 528.

Mitsch, W.J., L. Zhang, D. Marois, and K. Song. 2015. Protecting the Florida Everglades wetlands with wetlands: Can stormwater phosphorus be reduced to oligotrophic conditions? *Ecological Engineering* 80: 8–19.

News Service of Florida. 2018. Gov. Scott declares emergency over toxic algae outbreaks. *Tampa Bay Times*. Retrieved August 27, 2018.

Heaven to Hell and Back

From My Best Semester Ever to COVID Pandemic and Back

Professor Bill Mitsch receiving Professor Emeritus Award at Florida Gulf Coast University Commencement on December 18, 2022.

(Personal photo. Permission provided by W.J. Mitsch.)

DOI: 10.1201/9781003374619-9

9.1 Introduction

As the 2019 fall semester ended and our lab held its annual Christmas party at our usual location of Coconut Jack's in Bonita Springs in mid-December 2019 (**Figure 9.1**), I had a giddy sense of optimism that things were going so well with research funding, media attention (our studies were on countless national, Ohio, and Florida papers), and even the movie world, that I sensed that "reloading" with new grad students and projects was the right thing to do.

9.2 My best year ever

As depressing as the COVID-19 pandemic became starting in the second week of March 2020, the nine prior months of June 2019 through February 2020 had created some of our most productive and enjoyable experiences ever for any of my research labs. It included productive interviews and research in central Ohio during the 2019 summer (**Figure 9.2**) and extended international trips to **Denmark** (**Figure 9.3**), **Korea** (**Figures 9.4** and **9.5**), and **Estonia**

Figure 9.1 *Everglades Wetland Research Park (EWRP) post-fall semester seasonal celebration at Coconut Jack's in Bonita Springs, Florida, in mid-December 2019.*

(Personal photo. Permission provided by W.J. Mitsch.)

Figure 9.2 Bill Mitsch presents his Florida lab's research results at The Ohio State Fair, Columbus, Ohio, to WOSU radio, hosted by Ann Fisher on July 28, 2019.

(Personal photo. Permission provided by W.J. Mitsch.)

Figure 9.3 In June 2019, EWRP students, staff, and faculty gave four presentations at the 8th International Symposium on Wetlands and Pollutant Dynamics and Control (WETPOL), Aarhus, Denmark.

(Personal photo. Permission provided by W.J. Mitsch.)

Figure 9.4 Bill Mitsch presenting "Sustainably solving wetland loss and harmful algal blooms (HABs) with wetlands and wetlaculture" at the Society of Wetland Scientists-Asia Chapter and Korean Wetlands Society joint meeting on August 19, 2019, in Suncheon City, South Korea.

(Personal photo. Permission provided by W.J. Mitsch.)

Figure 9.5 Bill Mitsch visiting Professor Hojeong Kang's lab team in Seoul, South Korea on August 23, 2019. As a graduate student years earlier, Hojeong Kang spent several months doing Research at the Olentangy River Wetland Research Park that resulted in one of the first peer-reviewed papers coming for that Ohio wetland research facility (Kang et al., 1998). Professor Kang collaborated with Professor Mitsch and with their students on many other research projects related to wetland denitrification including work described in Song et al. (2010, 2012).

(Personal photo. Permission provided by W.J. Mitsch.)

Figure 9.6 In Tartu, Estonia, on December 7, 2019, with long-time collaborator Professor Ulo Mander from Tartu University. Professor Mander successfully nominated Professor Mitsch for a Doctorate honoris causa at the University of Tartu in 2010.

(**Figure 9.6**), as well as hosting Polish water scientists from Łódź, **Poland,** for a three-week Midwestern USA wetland, ecological engineering, and ecosystem tour (**Figures 9.7–9.11**) and sponsoring a Lake Erie-based workshop on solving harmful algal blooms (**Figure 9.12**).

9.3 A midwestern ecology tour for visiting Polish professors in summer 2019

Those of us who were doing research in Ohio during the summer 2019 hosted two Polish scientists, Professor Edyta Kiedrzyńska, Director of the European Regional Centre for Ecohydrology of the Polish Academy of Sciences and Department of Applied Ecology, and Professor Marcin Kiedrzyński, Laboratory of Plant Ecology and Adaptation, Department of Geobotany and Plant Ecology, University of Łódź, and their family for a three-week Midwest USA ecology tour, visiting Indiana, Michigan, and Ohio. The tours included peatlands and

Figure 9.7 *Polish family visiting the "Play Like a Champion" sign at the exit from the Notre Dame football team locker room at Notre Dame Stadium, Notre Dame, Indiana.*

(Personal photo. Permission provided by W.J. Mitsch.)

Figure 9.8 *FGCU wetlaculture experiment in Defiance County, Ohio, and upstream of Lake Erie. Wetlaculture (Mitsch et al., in press) is a landscape-scale integration of wetlands designed for the retention of nutrients (phosphorus and nitrogen) from polluted agricultural and urban runoff with systematic recycling of those nutrients to agriculture, horticulture, and/or forestry. The wetlaculture term comes from wetlands + agriculture.*

(Personal photo. Permission provided by W.J. Mitsch.)

Figure 9.9 *Indiana Dunes National Park on the southern shoreline of Lake Michigan, where climbing the dunes is facilitated by stairway boardwalks.*

(Personal photo. Permission provided by W.J. Mitsch.)

Figure 9.10 *Polish Professors Marcin Kiedrzyński and Edyta Kiedrzyńska toured several environmental laboratories at The Ohio State University during their visit.*

(Personal photo. Permission provided by W.J. Mitsch.)

Figure 9.11 *Polish Professors Marcin Kiedrzyński and Edyta Kiedrzyńska in one of several experimental wetlands at The Olentangy River Wetland Research Park at The Ohio State University in Columbus, Ohio. The wetland was curiously dry that day.*

(Personal photo. Permission provided by W.J. Mitsch.)

Figure 9.12 *Participants of a workshop we organized entitled, "Wetlands Mitigating Harming Algal Blooms," sponsored by The National Science Foundation, and held in Huron, Ohio, on August 3, 2019.*

(Personal photo. Permission provided by W.J. Mitsch.)

forests in Michigan but also cultural sites like the University of Notre Dame Football Stadium on July 30, 2019 (**Figure 9.7**); Florida Gulf Coast University (FGCU) wetlaculture experiments in northwest and central Ohio (**Figure 9.8**); and the Indiana Dunes National Park in northwest corner of Indiana on July 31, 2019 (**Figure 9.9**). The tour also included visits to several laboratories at The Ohio State University (**Figure 9.10**), including the internationally recognized Olentangy River Wetland Research Park at The Ohio State University (**Figure 9.11**).

The highlight of this Polish Midwestern tour was a conference co-sponsored by the Nastional Science Foundation on August 3, 2019, on the shoreline of Lake Erie in northern Ohio entitled, "Wetlands Mitigating Harming Algal Blooms" in Huron, Ohio (**Figure 9.12**). The two visiting Polish scientists were among 15 scientists making presentations to an audience of 100 during this all-day event. The program was designed for the public, not only for an academic audience, and received great press coverage. This was one of the best events our Everglades Wetland Research Park organized in its

10-year existence, and we repeated it again at our lab in Naples as part of our Moonlight on the Marsh lecture series in mid-February 2020; that version too was very well received. We had no way of knowing then that COVID-19 was then right around the corner.

9.4 A sea turtle movie success

A well-oiled research lab does not need to shut down because of a pandemic. There are students to support and grant deadlines as always. But nothing was the same after the COVID-19 pandemic manifested itself. But there were good things still going on. On September 18, 2019, we celebrated what turned out to be our only venture into the world of cinema with the release of an ocean conservation movie titled *Troubled Waters: A Turtle's Tale* (**Figure 9.13**). I delivered a few short narrations in the film, mostly about algal blooms such as red tide that were menacing the South Florida coastlines almost every year. Several more narrations were provided by film star Ted Danson. But the true stars were movie director Rory F. Fielding and the incredible underwater photography team, both of whom won Emmys. I was proud to be a bit-player in this PBS movie that was eventually shown all around much of the country during April 2020 for Earth Day celebrations.

I continued giving public lectures as talk of the pandemic continued (**Figures 9.14** and **9.15**). Whereas I had given 25 public lectures in 2019, I presented only 10 such lectures (and some were remote) in 2020, 15 in 2021, and only 11 in 2022.

9.5 The rest of the story

COVID-19 sneaked into our lives in early March 2020, just when our 2020 Moonlight on the Marsh lecture series was ending. Because we had added three new graduate students and a two-year, $1 million grant in the 2020–2021 academic year, we all knew that we had to figure out how to get these students and grants done while trying to avoid infection by COVID. It was not easy, nor were we always successful. I completed one Ph.D. student in our partnership program at University of South Florida in December 2021 and two M.S. grad students at FGCU in May 2022. Our $1 million grant from Florida Department of Environmental Protection (DEP) started in January 2021 and finally ended with a data collection and completion report on July 2022 (**Figure 9.16**). Having three successful theses/dissertation defenses and one final research report completed in less than two years was outstanding. Our students and staff contributed. We felt that we were in overdrive while continually dodging COVID outbreaks. There was no norm in those two years. It was then that I decided to retire at the end of September 2022 after 47 years as a professor.

Figure 9.13 *Bill Mitsch and movie director Rory Fielding celebrating the movie's opening night on September 18, 2019, for the ocean conservation movie titled* Troubled Waters: A Turtle's Tale, *in Fort Lauderdale, Florida. The movie was produced by WLR TV.*

(Personal photo. Permission provided by W.J. Mitsch.)

Figure 9.14 Bill Mitsch presenting "Investigating the causes and mitigation of harmful algal blooms in South Florida and The Laurentian Great Lakes" at The Water School, Florida Gulf Coast University, Fort Myers, Florida, on November 22, 2019.

(Personal photo. Permission provided by W.J. Mitsch.)

Figure 9.15 Bill Mitsch presents "Sustainably solving wetland loss and harmful algal blooms (HABs) with created wetlands and wetlaculture," on January 13, 2020, to the Seven Sisters of Southwest Florida luncheon, Pelican Isle Yacht Club, Naples, Florida.

(Personal photo. Permission provided by W.J. Mitsch.)

Figure 9.16 *Manual water sampling by my staff and graduate students in the two-year $1 million DEP contract titled "Monitoring, predicting, and controlling harmful algal blooms by buoy ultrasonic technology in a range of lakes in southwest Florida." Our team carried out the project with extraordinary success despite many supply-chain delays, legal and fiscal issues, and COVID always present in the background.*

(Personal photo. Permission provided by W.J. Mitsch.)

We ran an abbreviated Moonlight on the Marsh lecture series in March 2022 (**Figure 9.17**) with only two valedictory lectures. The first was by a long-term supporter of our labs at The Ohio State University and Florida Gulf Coast University, Dr. Bernard Master, presenting "Finding the rarest birds in the world." The second presentation was by me and titled, "What I fought for: Memoirs of an environmental professor," essentially the theme of this memoir.

Thanks to all the hundreds of great graduate students and staff who made my 47-year career possible!

Where are we now?

This book and my lecture in March 2022 on "what I fought for" will not be the final word for me even after almost a half-century in the academic world. Many friends and former students came down to Naples, Florida, in March 2022, to listen to me one more time (**Figure 9.18**).

Writing this book has been cathartic for my soul, because as I wrote these chapters, I sometimes questioned the who or why of some of my decisions, but I also recognized that I was a fighter for the environment and always will be. The message that should go to my younger readers and students who

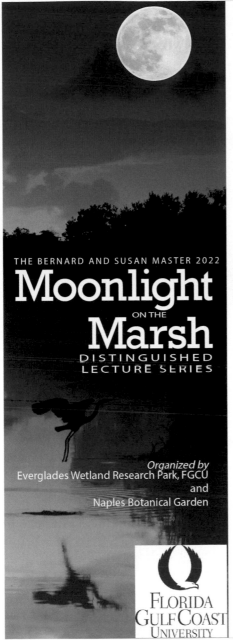

Figure 9.17 *Our 2022 Special Valedictory "Moonlight on the Marsh" lecture series with an abbreviated schedule of lectures by Dr. Bernard Master and me due to COVID risks. My presentation on March 24, 2022, was my retirement presentation at FGCU to a sold-out and appreciative audience (see* **Figure 9.18***).*

(Personal photo. Permission provided by W.J. Mitsch.)

Figure 9.18 *After my Valedictory "Moonlight on the Marsh" lecture on March 24, 2022, many of my grad students and advisory committee members came up for a group photo shoot. I hope many of you see yourself in this photo when the book gets published!*

(Personal photo. Permission provided by W.J. Mitsch.)

might buy this book is to find out what floats your boat before you sail on. If you are interested as I was in defining a real environmental science that is indisputable truth, it may not ever happen easily. Working at the edge between science and policy will always be difficult, but it needs to happen so we can protect the natural environment for its beauty and diversity but also for all the ecosystem services nature gives to us freely.

I am hoping that this book will show that, in the real world, accomplishing important things is never easy nor are we ever done. For example, Chapter 5 illustrates how politicians continue to dispute the values of nature and wetlands by redefining "waters of the United States" every 5 to 10 years or so. We were so proud to think that we saved hundreds of thousands of hectares/ acres of wetlands in 1995 by stopping Congress from defining wetlands in 1995 only to find out that the same conflicts came back in 2020.

Another advancement that developed during these pandemic years is the possibility of using Zoom and Microsoft Teams to present remotely talks that

Table 9.1 Remote Presentations Given by Bill Mitsch in the Past Year

- **November 2, 2022**—"Future of Wetland Research in China and at Northeast Normal University in Changchun, China," 20th Anniversity Celebration of Key Laboratory of Wetland Ecology at Northeast Normal University, Changchun, China and Bonita Springs, Florida, USA.
- **October 27, 2022**—"The role of Wetlands in the global carbon cycle and climate change," XIII Simposio Internacional del Carbono en México, Mexico City, Mexico.
- **June 8, 2022**—"Nature-based solutions and ecological engineering: Our best hopes for restoring lakes, rivers, and estuaries and protecting human health," remote presentation, ecohydrology for water security session at the 5th International Symposium on Healthy Rivers and Sustainable Water, Warsaw, Poland.
- **April 15, 2022**—"Nature-based solutions: Our best hope for restoring lakes, rivers, and estuaries and protecting public health," remote presentation, Ohio Environmental Health Association Annual Meeting, Dublin, Ohio.
- **January 8, 2022**—"Nature-based solutions: Our best hope for restoring the Everglades," Everglades Coalition, Nature-Based Solutions Panel Discussion, Hawks Cay Resort in Duck Key, Florida.
- **December 1, 2021**—"50 years since the first Earth Day: Bill Mitsch's 50-year Career Dealing with Troubled Waters, Saving Wetlands and Developing Ecological Engineering," American Ecological Engineering Society, North Carolina State University

otherwise would require days of travel and major costs to do. I believe remote presentations are going to be the future for many national and international meetings and presentations. In 2021–2022, I gave six remote presentations listed here in **Table 9.1**, including international presentations in Poland, Mexico, and China that would have been prohibitively expensive to me and my hosts to be presented live.

My advice for those of you who are interested in protecting the environment is to understand that science can only describe what is going on to a point. Afterward, we need to appreciate that policy is incomplete and sometimes we may need to recalibrate our views. But having an appreciation of the natural world is one of the best starting points for getting involved. Learn about ecosystems, how they work, and the services that they provide to all of us and only then you can fight for what you think is right in this wonderful Planet Earth that we should be privileged to protect and manage.

References

Kang, H., C. Freeman, D. Lee, and W.J. Mitsch. 1998. Enzyme activities in constructed wetlands: Implication for water quality amelioration. *Hydrobiologia* 368: 231–235.

Mitsch, W.J., B. B. Jiang, S. Miller, K. Boutin, L. Zhang, A. Wilson, and B. Bakshi. 2022. Wetlaculture: Solving harmful algal blooms with a sustainable wetland/agricultural landscape. In B. R. Bakshi (ed.), *Engineering and Ecosystems: Seeking Synergies for a Nature-Positive World*. Springer.

Song, K, H. Kang, L, Zhang, and W.J. Mitsch. 2012. Seasonal and spatial variations of denitrification and denitrifying community structure in created wetlands. *Ecological Engineering* 38: 130–134.

Song, K., S.H. Lee, W.J. Mitsch, and H. Kang. 2010. Different responses of denitrification rates and denitrifying bacterial communities to hydrologic pulsing in created wetlands. *Soil Biology & Biochemistry* 42: 1721–1727.

Index